Business Power and Conflict in
International Environmental Politics

Also by Robert Falkner

The International Politics of Genetically Modified Food: Diplomacy, Trade and Law (edited)

The Cartagena Protocol on Biosafety: Reconciling Trade in Biotechnology with Environment and Development (co-edited with Christoph Bail and Helen Marquard)

Business Power and Conflict in International Environmental Politics

Robert Falkner

London School of Economics and Political Science, UK

First published 2008 by
PALGRAVE MACMILLAN
Houndmills, Basingstoke, Hampshire RG21 6XS and
175 Fifth Avenue, New York, N.Y. 10010
Companies and representatives throughout the world

PALGRAVE MACMILLAN is the global academic imprint of the Palgrave
Macmillan division of St. Martin's Press, LLC and of Palgrave Macmillan Ltd.
Macmillan® is a registered trademark in the United States, United Kingdom
and other countries. Palgrave is a registered trademark in the European
Union and other countries.

ISBN-13: 978–0–230–57252–2   hardback
ISBN-10: 0–230–57252–9   hardback

This book is printed on paper suitable for recycling and made from fully
managed and sustained forest sources. Logging, pulping and manufacturing
processes are expected to conform to the environmental regulations of the
country of origin.

A catalogue record for this book is available from the British Library.

A catalog record for this book is available from the Library of Congress.

10 9 8 7 6 5 4 3 2 1
16 15 14 13 12 11 10 09 08 07

Printed and bound in Great Britain by
Antony Rowe Ltd, Chippenham and Eastbourne

To Kishwer

Contents

Preface

This book was written with two audiences in mind.

The first audience comprises students and practitioners of diplomacy who see international politics dominated by states, and particularly by the great powers. In this view of inter-state relations – what Karl Marx mockingly described as the 'high-sounding drama of princes and states' – the only actors that can create lasting international institutions and provide authoritative solutions for global problems such as environmental degradation are states. Against this view, I argue that global power is more widely dispersed among different types of international actors, state and nonstate, and that if we wish to gain a better understanding of the international politics of the environment, we need to pay closer attention to how business shapes outcomes within and beyond the field of environmental diplomacy.

The second audience consists of all those, including Marxists, who take economic factors in international politics seriously, but who end up overstating the power and influence of business. By showing the connections between the underlying capitalist order and international political outcomes, they bring to light the often hidden dimensions of global business power. Again, against this view, I argue that business interests and strategies are diverse and potentially conflicting, and that the potential for business conflict serves to limit overall business power and influence in international environmental politics.

This book, therefore, contains two broad arguments about the international role of business. It seeks to highlight the many ways in which corporate actors operate from a privileged position vis-à-vis states and NGOs, when it comes to setting global environmental standards and implementing environmental agreements. Due to their role in directing investment and technological innovation, companies can set the parameters of what is politically achievable in the international politics of environmental protection. Yet, business actors rarely act in unison, and references to an underlying business or class interest fail to explain the competitive dynamics that characterize business involvement in international politics. Business conflict opens up political space for other actors – states, international organizations and social movements – to press for global change. The bond that holds these two arguments together is the neo-pluralist perspective on business power that is developed in more detail in chapter 2. Its basic message is this: business actors are in a powerful, even privileged, position internationally, but they don't always get their way. The process of international environmental politics is more fluid and open-ended than it may seem at first sight, and while business interests sometimes predominate they don't control the global environmental agenda.

The two perspectives against which the neo-pluralist argument is developed – statism and structuralism – have long occupied polar ends on the wide theoretical spectrum in the study of international relations and international political economy. As with all such 'isms', they represent highly stylized views that have been critiqued before, but they frequently surface in scholarly and popular debates on globalization and global governance. Their attraction lies in the fact that they offer a neat, coherent and powerful account of the forces that drive international politics. Admittedly, the alternative perspective I offer in this book yields a messier picture. But it is one, I hope, that better captures the complex and open nature of international environmental politics, and that shows where opportunities lie for agents of change to bring about a greener future.

This book is based on several years of research on international environmental politics. Academic research is usually a solitary occupation, but rarely progresses without the assistance of others. I have been fortunate to receive generous support, constructive criticism and, above all, inspiration from a number of teachers, colleagues and friends. I would like to thank them all, and apologize to those whom I have forgotten to include.

Andrew Hurrell guided me through my doctoral studies at Nuffield College, Oxford. His enthusiasm for my project, on the role of global firms in ozone politics, helped me to get started in this area, and without his encouragement, advice and patience I would not have been able to complete it. My college supervisors at Nuffield, Yuen Foong Khong, Joachim Jens Hesse and Jeremy Richardson, were always there to give feedback and moral support. Gillian Peele and Elizabeth Frazer eased me into my first teaching positions, at Lady Margaret Hall and at New College. Vivienne Jabri, Jef Huysmans and Mervyn Frost helped me to settle in at the University of Kent's London Centre of International Relations and gave me the necessary space to continue with this research. My time at Essex University was too short to fully benefit from the expertise of Albert Weale and Hugh Ward, but in their own way they have inspired me to continue on this path. The London School of Economics eventually became my institutional and intellectual home, and I have benefited greatly from working with colleagues in International Relations and other departments. For many years, I have been fortunate to carry out research on international environmental policy at Chatham House, where under the leadership of Duncan Brack and Richard Tarasofsky, the Energy, Environment and Development Programme has been a place for research and reflection, and for exposing academic ideas to the 'real world'.

Over the years, several colleagues and friends have guided me in my research or commented on earlier writings that have made it into this book. My thanks go to Stephen Anderson, Christoph Bail, Tanja Börzel, Duncan Brack, Jennifer Clapp, Robyn Eckersley, Manfred Elsig, Doris Fuchs, Aarti Gupta, Andrea Liese, David Levy, Helen Marquard, Tom Morehouse, Peter Newell, Kalypso Nicolaïdis, Matthew Paterson, Richard Tarasofsky, David Vogel, John Vogler, Andrew Walter and Ngaire Woods.

In the course of writing this book, I have received the support of a number of institutions. I was able to conduct several collaborative research projects under the auspices of the Energy, Environment and Development Programme at Chatham House. The London School of Economics gave generous research leave to free up time to work on this book, and the Minda de Gunzburg Center for European Studies at Harvard provided a convivial environment in which to put the finishing touches to it. I gratefully acknowledge the financial support I have received over the years to conduct archival research, interview decision-makers and attend UN conferences. Thanks go to the Deutscher Akademischer Austausch Dienst; the Citibank Scholarship Programme, Cyril Foster Fund and Goodhart Fund at Oxford; the German Historical Institute in Washington, DC; the Rockefeller Foundation; and the John D. and Catherine T. MacArthur Foundation.

I am grateful to Philippa Grand at Palgrave Macmillan, who gave her enthusiastic support to this project, and to Hazel Woodbridge, who saw it through to its completion.

Thanks are due to MIT Press and Taylor & Francis for allowing me to use previously published material in this book. Portions of chapter 3 draw on 'The Business of Ozone Layer Protection: Corporate Power in Regime Evolution', in David L. Levy and Peter J. Newell (eds) *The Business of Global Environmental Governance* (MIT Press, 2005), pp. 105–34; and parts of chapter 5 build on 'The Political Economy of "Normative Power" Europe: EU Environmental Leadership in International Biotechnology Regulation', *Journal of European Public Policy* 14(4) 2007: 507–26.

Special thanks go to my family. To my parents, Barbara and Max, for their loving support all those years ago and today. To my daughter Sophia, who offers a sense of perspective and has always been a welcome source of distraction. And, last but not least, to my wife Kishwer, who was there from the start of this journey, and without whom I could not have completed it. I dedicate this book to her.

Robert Falkner

List of Abbreviations

AGBM	Ad Hoc Group on the Berlin Mandate
AGBM-n	n^{th} meeting of the Ad Hoc Group on the Berlin Mandate
AIA	advance informed agreement
AOSIS	Alliance of Small Island States
BCSD	Business Council for Sustainable Development
BELC	Business Environmental Leadership Council
BIO	Biotechnology Industry Organization
BP	British Petroleum
BSWG	Open-Ended Ad Hoc Biosafety Working Group
BSWG-n	n^{th} meeting of the Biosafety Working Group
CAFE	Corporate Average Fuel Economy
CBD	Convention on Biological Diversity
CCP	Cities for Climate Protection
CCX	Chicago Climate Exchange
CDM	Clean Development Mechanism
CDP	Carbon Disclosure Project
CFCs	chlorofluorocarbons
CMA	Chemical Manufacturers Association
CO_2	carbon dioxide
COP	Conference of the Parties
COP-n	nth meeting of the Conference of the Parties
COP/MOP	Conference of the Parties serving as Meeting of the Parties
COP/MOP-n	nth meeting of the Conference of the Parties serving as Meeting of the Parties
CSR	corporate social responsibility
EPA	Environmental Protection Agency
EU	European Union
ExCOP	extraordinary meeting of the Conference of the Parties
GCC	Global Climate Coalition
GDR	German Democratic Republic
GHG	greenhouse gas
GIC	Global Industry Coalition
GM	genetically modified
GMO	genetically modified organism

HCFCs	hydrochlorofluorocarbons
HFCs	hydrofluorocarbons
ICC	International Chamber of Commerce
ICCP	International Climate Change Partnership; and Intergovernmental Committee for the Cartagena Protocol
ICI	Imperial Chemical Industries
INC	Intergovernmental Negotiating Committee
IPCC	Intergovernmental Panel on Climate Change
IPE	International Political Economy
JI	Joint Implementation
JUSSCANNZ	group of countries comprising Japan, United States, Switzerland, Canada, Australia, Norway and New Zealand
LMO	living modified organism
LMO-FFPs	living modified organisms for direct use as food, feed, or for processing
MEA	multilateral environmental agreement
MNC	multinational corporation
NAFTA	North American Free Trade Agreement
NGO	non-governmental organization
NIH	National Institutes of Health
ODS	ozone-depleting substances
PR	public relations
OECD	Organisation for Economic Cooperation and Development
OPEC	Organization of Petroleum Exporting Countries
R&D	research and development
SAGB	Senior Advisory Group for Biotechnology
SPS	Sanitary and Phytosanitary Measures (WTO Agreement)
UN	United Nations
UNCED	United Nations Conference on Environment and Development (also known as the 'Earth Summit')
UNEP	United Nations Environment Programme
UNFCCC	United Nations Framework Convention on Climate Change
UNICE	Union of Industrial and Employers' Confederations of Europe
US	United States
WBCSD	World Business Council for Sustainable Development

WICEM	World Industry Conference on Environmental Management
WTO	World Trade Organization
WWF	Worldwide Fund for Nature ('World Wildlife Fund' in the US and Canada)

I: Introduction

1
Global Firms in International Environmental Politics

Global environmental crises such as climate change and the loss of biological diversity pose a major challenge to the international system. Organizing a collective response to environmental degradation has proved difficult in the past. Diverging national interests and weak international institutions have held back efforts to protect the global environment. But even when a global political consensus can be found, myriad social and economic forces may end up obstructing effective environmental governance. The globalization of economic production, consumption and exchange is fuelling environmental destruction while at the same time complicating the search for political solutions (Hurrell and Kingsbury, 1992).

The contradictions between global economic and environmental trends have become evident as the global ecological crisis unfolds and political responses remain ineffective. Despite a growing recognition that carbon-based energy sources are a key source of global warming, demand for fossil fuels continues to outpace political efforts to mitigate climate change (Houghton, 2004). The loss of tropical forests, and with it of biological diversity, is in part caused by international demand for timber, but the international regulation of this trade is fragmented and weak (Dauvergne, 2001). And efforts to protect biodiversity from the unintended release of genetically modified organisms run up against the uncontrolled spread of biotech crops through the global food chain and international aid channels (Clapp, 2006). To many observers, the deepening ecological crisis suggests that the global economy is out of control (Hertz, 2001; Monbiot, 2000).

When speaking of the economic forces that obstruct global environmental policy, many refer to capitalism, or industrialism, or

3

simply globalization. What remains hidden in such overarching concepts is the role that specific economic actors play, most notably corporations. The agency of business is of critical importance here, for it is global firms that operate at the nexus between the global economy and international politics. Business actors translate abstract economic forces into concrete, observable, facts of political life. Their dual role as economic *and* political actors makes them pivotal players in the search for global environmental solutions. To arrive at a better understanding of the dynamics and limits of international environmental politics, therefore, we must focus on the power and influence of business.

Today, corporations are more actively involved in international environmental politics than ever before, though their role has changed dramatically in recent years. In the early days of global environmentalism, during the 1960s and 1970s, most business leaders viewed environmental ideas with suspicion and hostility. Environmental and economic objectives appeared to be mutually exclusive, and a straightforward trade-off seemed to exist between environmental protection and corporate profitability. In this early phase, corporate involvement in international environmental politics was limited to occasional, and largely reactive, interventions to prevent burdensome regulations. During the 1980s, environmental problems did not disappear from the international agenda, as many business leaders had hoped. Nor did environmentalism. A number of major industrial accidents underlined the severity of the ecological crisis, and the corporate sector started slowly to change tack. A growing number of corporations began to integrate environmental objectives into their business operations and some developed more progressive political strategies. The 1987 Montreal Protocol to protect the ozone layer marked a turning point in this process, as it was one of the first international environmental treaties that attracted the support of the very industry that had caused the problem of ozone depletion in the first place (Hoffman, 1997).

What is striking about this changing field of corporate environmental strategy is the diverse set of business interests and approaches that has evolved. Some corporations continue to oppose international environmental regulations as an unwarranted burden on their operations, while others openly support higher international regulatory standards. Some corporate leaders remain hostile to the idea that environmental concerns should limit their economic freedom, while others have begun to speak of the need to reconcile the goals of economic development and ecological sustainability. Undoubtedly, some of the so-called 'greening' of industry is little more than empty rhetoric. But it signifies a more profound and potentially lasting trend that has had a notable impact on international politics. Corporations can now be found on different sides of global debates, arguing for or against environmental regulation. To make sense of the complex forces that shape international environmental politics, we need to look more closely at corporate involvement in international politics. What is needed is a better understanding of business power, of its nature and its sources, its impact on international outcomes as well as its limits. This book aims to contribute to a better understanding of how business has become part and parcel of the international politics of environmental protection. It develops an analytical framework that allows us to examine the often contradictory roles that corporations play and argues that a neo-pluralist perspective is best placed to make sense of the changing dynamics that characterize this policy field. The neo-pluralist argument will be developed in greater detail in chapter 2.

Business and the environment: towards engagement

Corporations have long been concerned with environmental and workplace safety issues, but it is only recently that they have had to engage with international environmental policy-making in a more systematic way. The rise of the modern environmental movement in the 1960s and the emergence of a global environmental agenda in the 1970s created pressures for change that were initially met with business responses ranging from indifference to denial and resistance. Corporate leaders either questioned the scientific evidence of environmentalists' demands or rejected them on the grounds that they would harm economic growth, innovation and employment. Against the background of a worsening global economic climate in the 1970s and early 1980s, business opposition to ever stricter environmental regulation seemed to pay off as environmental concerns took a back seat again in most industrialized countries (Hoffman, 1997: 3–4).

But if business leaders held any hopes that environmentalism would ebb away, these were in vain. The first United Nations (UN) environment conference in Stockholm in 1972 was followed by some important institutional innovations that helped to entrench environmental issues on the global political agenda. The United Nations Environment Programme (UNEP) was established in 1973 and became a major site for developing international environmental instruments and treaties. At national level, most industrialized countries institutionalized environmental policy by creating specialist ministries and agencies that gradually managed to expand the scope of environmental regulation. They were able to sustain this trend not least because of a string of high profile industrial accidents, which further eroded public confidence in industry self-regulation. The chemical spill in Seveso, Italy (1976), the explosion at a Union Carbide plant in Bhopal, India (1984), and the meltdown at the Chernobyl nuclear power plant in Ukraine (1986) all contributed to the growing sense that the untrammelled growth of industrialism produced unacceptable risks to the environment and human health (Elliott, 2004: 12–13).

The persistence of the environmental agenda made a change in corporate response and strategy necessary. As environmental regulations covered ever more aspects of corporate activities and compliance costs rose, a reactive and largely obstructionist business approach would no longer do. Of course, industrial firms could in theory move to countries with lower environmental standards. While there is some anecdotal evidence for such regulation-induced 'industrial flight', recent research suggests that it has been a more isolated phenomenon than is commonly assumed (Neumayer, 2001: chapter 3). There are also important reasons why companies seek instead to adapt to a changing regulatory field and engage with the environmental agenda. For one, relocating manufacturing operations abroad is not an option for those industries that need to operate in close proximity to skilled labour or consumer markets in the industrialized world. This is the case with, for example, research-intensive industries such as biotechnology and pharmaceutical companies. Moreover, environmental costs are often only a small part of the overall cost structure of production, and technologically advanced firms may find it more cost-effective to reduce environmentally harmful activities in order to limit regulatory costs.

As corporations started to take the environmental challenge more seriously from the 1980s onwards, they also began to develop internationally coordinated responses. Initiated by the International Chamber of Commerce (ICC) and UNEP, the first ever World Industry Conference on Environmental Management (WICEM) took place in 1984 and brought together over 500 international business and government representatives. The meeting led to a follow-up event in 1991, WICEM II, which saw the launch of the *Business Charter for Sustainable Development*, a set of 16 principles for environmental management that have been influential in the development of voluntary business codes across different sectors. Parallel efforts to develop environmental management systems came under way at around the same time. In 1991, the International Organization for Standardization set up a Strategic Advisory Group on the Environment in order to explore the need for internationally harmonized standards dealing with sustainability issues. These discussions gathered pace after the 1992 UN Conference on Environment and Development and led to the adoption of the ISO 14000 series, a global standard for environmental management that has been widely adopted especially by multinational corporations (MNCs) (Prakash and Potoski, 2006).

The change in business attitudes became evident in the run-up to the 1992 UN Conference on Environment and Development (UNCED). The so-called Rio 'Earth Summit' sparked considerable interest within the business sector, and saw an unprecedented level of business involvement in the preparations and negotiations. Representing a broad coalition of industry interests, Swiss industrialist Stefan Schmidheiny served as special advisor to the secretary-general of UNCED, Maurice Strong, thus allowing business to play a uniquely high-profile role in the conference proceedings. Schmidheiny had gathered 48 business leaders from the chemical industry to electronics manufacturing and the oil sector to form the Business Council for Sustainable Development (BCSD) (Schmidheiny and BCSD, 1992), a body now known as the World Business Council for Sustainable Development (WBCSD). Their hope was to inject a more business-friendly approach into international environmental politics and to legitimize business as a partner in the search for global solutions, something that was eventually recognized in chapter 30 of Agenda 21, UNCED's main plan of action.

Corporations also came to involve themselves more actively in other environmental negotiations. The talks on a global regime to combat ozone layer depletion proved to be a watershed event in this regard. Not only did key industry groups participate as observers in the talks, but some of them also came out in support of international restrictions on ozone-depleting chemicals, thus breaking with a long tradition of business opposition to international environmental regulation. The case of the 1987 Montreal Protocol on ozone layer depletion was widely hailed as a model for business–government relations. That greater industry involvement could turn out to be both a help and hindrance in the creation of strong environmental regimes became clear in the subsequent negotiations on a climate change treaty.

A more recent development has been the emergence of new types of environmental governance mechanisms outside the realm of state-centric international politics, so-called 'private environmental governance' (Falkner, 2003). On numerous occasions, businesses, environmental non-governmental organizations (NGOs) and international organizations have cooperated to create norms, rules and mechanisms for environmentally friendly corporate behaviour. Large MNCs especially have engaged in this form of private governance, to respond to environmental concerns and to protect their reputation as well as to avert potentially more stringent public regulation. They have created codes of conduct, certification schemes and other mechanisms that govern economic transactions in areas such as forestry, project finance and environmental reporting (Wright and Rwabizambuga, 2006; Pattberg, 2007).

Underpinning these changes has been a growing recognition that corporations have a responsibility to avert harm to society and the environment, beyond what existing laws and regulations stipulate. This movement towards corporate social responsibility (CSR) has been in the making for several decades but gained particular momentum in the 1990s, as more and more global firms designed their own CSR policies and signed up to industry-wide codes of conduct (Vogel, 2005). Many CSR schemes have been criticized for being vague, unenforceable, and ultimately ineffective. But the fact remains that the rise of the CSR agenda has pushed global environmental concerns to the fore of debates on global corporate strategy and has created a broader societal expectation that corporations develop proactive strategies to minimize their negative environmental impacts. Moreover, CSR principles are slowly being strengthened through codification or adoption by governments who seek to create a level playing field for all companies in respect to accountability, transparency and responsibility criteria.

To sum up this brief historical overview, within the last three decades, public and business discourses on environment have shifted from an 'economy-versus-ecology' perspective to greater integration and engagement of economic and environmental agendas. Individual corporations may continue to oppose specific environmental measures, but most will acknowledge the need for global environmental governance and an integration of environmental objectives into corporate strategy. This poses new challenges for the study of international environmental politics and the role of business in international relations more generally. Analysts need to take into account the changing nature of business involvement and the growing diversity of corporate strategies. A brief overview of the different roles that firms play in this policy field will illustrate the complexity of international environmental politics.

Corporate roles in international environmental politics

Corporations have engaged with the international politics of environment in a number of ways. Broadly speaking, we can distinguish four different roles that they play, both within and outside the realm of international environmental diplomacy.

Lobbying in international negotiations: Corporations have become increasingly active as interest groups in international negotiations. As the number and scope of multilateral environmental agreements (MEAs) has expanded over the last three decades, so has the number of business lobbyists focusing on international policy-making. This form of business involvement is the most widely researched in international relations, and has been documented across a range of policy issues (Sell, 2003; Levy and Prakash, 2003). We have a fairly good understanding of how sectoral differences shape corporate preferences with regard to foreign economic policy (Frieden, 1988), and models exist to examine the multiple levels of interaction between domestic interest groups, state actors and the international bargaining process (Evans et al., 1993). In environmental politics, corporations seek to influence the outcome of negotiations for a variety of reasons: to prevent the creation of international regulations that would harm them; to make international regulations more business-friendly in cases where some form of regulation is either desirable or inevitable; and to shape the specific nature of regulations to create business opportunities, for example in cases where regulation creates new markets or gives competitive benefits to technological leaders. The differential effects of regulation thus produce winners and losers and will shape corporate preferences and lobbying efforts accordingly (DeSombre, 2000: 34–9).

Implementation and technological innovation: Once an international environmental agreement has been adopted, corporations play an important part in the implementation phase. Their decisions on how to respond to new regulatory requirements will have an important impact on the effectiveness of international regimes. Corporate decisions on investment, product and process changes and technological innovation all take on a political dimension as they determine how regulations translate into economic change. Regulation, therefore, is not simply 'technology-forcing' but interacts with, and crucially depends on, choices made in the corporate sector. Moreover, corporations possess considerable room for manoeuvre vis-à-vis domestic regulatory agencies and often manage to weaken the impact of regulations, a phenomenon that is known as 'regulatory capture'. Corporations thus shape the outcome of international regime-building efforts indirectly at the implementation stage, by producing political and technological feedback mechanisms that enhance or limit the effectiveness of regimes (Falkner, 2005).

Shaping public discourses: International environmental politics takes place within a wider ideational context. Environmental norms, ideas and discourses set the social context in which environmental policy is developed and political choices are made (Litfin, 1994; Hajer, 1995). Environmental discourses shape public perceptions of environmental problems and influence which problems make it onto the political agenda, and normative frameworks guide the policy process and inform actors' interests and identities. It should therefore come as no surprise that corporations have also sought to influence environmental discourses and inject a more business-friendly perspective on regulatory politics (Fuchs, 2005: 146–9). Business actors have sought to gain public legitimacy by establishing themselves as providers of environmental solutions and have pushed for a market- and growth-friendly interpretation of sustainable development.

Private norm and rule-setting: Corporations have always relied on self-regulation to address issues of public concern, but have more recently established international mechanisms of private governance. Spurned on by environmental activist and consumer groups, many global corporations have set up codes of conduct and created global CSR schemes, often with other nonstate actors and international organizations (e.g. UN Global Compact). This phenomenon, which has also become known as 'private regimes' (Haufler, 1995) or 'private authority' (Cutler et al., 1999), is now a mainstay of global environmental policy-making, covering issues from international timber trade to fisheries protection and fresh water resources. The rise of private governance mechanisms has further strengthened and legitimized the role of business in the global politics of the environment.

Having thus charted corporate involvement in international environmental politics and its different manifestations, we now turn to the question that is central to this study: how powerful are corporations in international environmental politics? How much influence do they yield, and to what extent are they able to shape the emerging global governance architecture for environment? To answer these questions, we will need to engage more deeply with the concept of power, and specifically of business power (see chapter 2). Questions of power have always been central to International Relations, but only more recently the focus has shifted to nonstate actors such as business. The following brief review of this field of enquiry will lay out the context in which theoretical perspectives on business power have evolved.

Analytical perspectives on business power

The study of business power has become a central concern in international political economy (May, 2006). Economic globalization and the expansion of transnational relations have been the driving force behind this trend. This is in contrast to the origins of International Relations as a discipline with a distinctive state-centric orientation. In the then dominant tradition of realism, corporations were seen as marginal or subordinate actors in an international system populated by states and driven by state interests. By the 1960s and 1970s, however, the ongoing transformations in the post-Second World War international order made it necessary to widen the analytical gaze and give nonstate actors proper theoretical and empirical attention (Keohane and Nye, 1971). This has since grown into a well-established research area often referred to as transnationalism (Risse-Kappen, 1995).

The conclusion that nonstate actors such as corporations play a more visible role in international relations is now widely accepted, but debate continues on just how much they matter or, in other words, whether they make a difference. Popular debates on globalization frequently refer to the size of multinational corporations, particularly their trading volume and market capitalization, as proxy indicators of how powerful firms have become. But such rough indicators are an unreliable guide to measuring the impact that corporations have in international politics (Fuchs, 2005: 75–7), and while they give a first-cut sense of their economic power, these often provide misleading accounts of their political power. The very question of how MNCs' economic power translates into political power is at the heart of all discussions of their role and influence in international affairs. What is needed, therefore, is a better understanding of the different dimensions of corporate power, and how it gives rise to influence in relations with states and other nonstate actors in an international context.

In political science, a standard approach to the study of business power has been to treat business actors as interest groups that seek to influence policy outcomes within the state. Just like other interest groups, such as trade unions, consumer associations, activist groups and religious communities, corporate actors possess a specific set of resources that they can use to shape public policy debates and influence decision-making processes. Their overall influence depends on the relative strength of their power resources and the political strategies they employ. In this pluralist perspective, interest groups compete for political influence, and policy processes are in principle open-ended and non-deterministic.

On certain issues and in certain policy fields, some interest groups may be more influential than others but, overall and especially in the long run, competition between the different groups will create some kind of balance of interests that prevents one group from dominating a policy field altogether. This pluralist framework, originally developed in the context of American politics, can be transferred to international relations, too, where nonstate actors equally compete for influence in international decision-making.

The pluralist understanding of interest groups has provoked criticism in both political science and international relations. International Political Economy (IPE) scholars, in particular, have questioned the implicit assumption that business is an interest group just like any other, and that the policy process offers a level playing field for all interest groups. Instead, they have argued that business is in a privileged position because of its critical role in the economy, as a provider of employment, as the source of economic growth and as a stimulus for technological innovation. Similar questions can be raised about the role that corporations play in international environmental politics. Given that most international environmental agreements and regulations seek to bring about a change in corporate behaviour (e.g. reduce pollution, replace hazardous products or change risky processes), the business sector is likely to be in a strong position to block stringent regulations or weaken international rules once they have been adopted. Corporations control key resources – financial, technological and organizational – that play a critical role in determining the effectiveness of international environmental regimes. Their central role in deciding on investment and innovation is bound to give them a prominent position in the international political process, one that may eclipse that of other actors.

This critique of pluralism has led to alternative IPE perspectives that are based on a structuralist understanding of power. In these views, material forces that exist in the international political economy systematically favour one type of actor, business, over other actors. In the broadest sense, structuralists claim that ownership of capital gives rise to a diffuse but effective form of power that controls the context in which state actors reach decisions on specific policy questions, including by preventing certain issues from reaching the political agenda. Applied to the case of international environmental politics, these approaches provide important correctives to the pluralistic view of business power. For example, analysts have highlighted the importance of fossil fuel-based energy to the functioning of industrial systems, which in turn allows the oil and coal industry to exercise a kind of veto power over stringent international measures to combat global warming (Newell and Paterson, 1998). This line of reasoning has helped to focus scholars' attention on the pervasive role of business in shaping environmental policy outcomes and limiting the scope for international regulation overall. But structuralist explanations of business power run the risk of reducing international policy processes and outcomes to manifestations of immovable structures of business power at the cost of political agency. Despite theoretical advances in structuralist approaches that seek to overcome the deterministic tendency of historical materialism (Levy and Newell, 2005), structuralist accounts still struggle to deal with the observable variation in policy outcomes and the gradual emergence and growth of a global environmental agenda.

To stay with the example of climate change, the structural power of the fossil fuel industry explains some, though not all, difficulties encountered in the creation of the climate regime, but not the differences in corporate strategy and governmental positions between Europe and North America, nor the successful conclusion of the Kyoto Protocol negotiations in 1997. The very fact that the protocol has now come into force – a development that only a few years ago seemed highly doubtful – suggests that global environmental politics is a more open-ended and pluralistic process in which political agency can overcome structural impediments.

What is needed, therefore, is an understanding of business power as a multi-faceted and multi-dimensional phenomenon, and an analytical perspective that is reflective of the political agency of firms as well as their structural power. This perspective would need to be sensitive to the privileged position that business enjoys while acknowledging the diversity of business interests and the potential for conflict within the business community over matters of political strategy. As will be argued in chapter 2, neo-pluralism provides this perspective.

At the heart of the neo-pluralist understanding of business in international politics is the recognition that, while business may be in a privileged position, its power and influence over international outcomes need to be established in the context of specific issue areas and fields of activity. Countervailing forces, which are located in the international and transnational spheres, limit corporate influence as do divisions within the business sector itself. Indeed, the potential for what has become known as 'business conflict' (Nowell, 1996), that is the cleavages between different firms and industrial sectors with regard to international politics, prevents an understanding of business actors as belonging to a monolithic bloc or representing a class interest. Neo-pluralists hold that the unity of business interests and strategy are a matter of empirical study, not theoretical conjecture. Likewise, the existence of structural business power needs to be established empirically and cannot prejudge the question of how powerful business actors are in specific international contexts. In other words, not all business actors are engaged in international politics; not all of those that are share the same interest; and not all of those that seek to influence international politics succeed.

Overview of the book

This book focuses on three high-profile cases in recent international environmental politics: ozone depletion, climate change and agricultural biotechnology. Chapters 3, 4 and 5 provide detailed case studies of the involvement of business actors in the international political process in these environmental policy areas. The focus is on international regime creation, especially on the negotiations that have led to the creation of the three international treaties that are at the centre of international governance in these areas: the 1987 Montreal Protocol on substances that deplete the ozone layer, the 1997 Kyoto Protocol on climate change, and the 2000 Cartagena Protocol on Biosafety. Where appropriate, the implementation of these treaties is also discussed, as are wider efforts to create environmental governance mechanisms around and outside the realm of state-centric environmental diplomacy.

The three chapters on ozone depletion, climate change and agricultural biotechnology were written in a chronological order, tracing the evolution of international policy-making and linking it with the changing dynamics of business power and business conflict. This, it is hoped, will allow readers to gain an overview of the main issues and developments in these global policy fields, so that each chapter can be read on its own, as a case study in international environmental policy-making. Given the focus on the role of business, it was not possible to do justice to the full complexity of each case, or to cover all major issues and actors involved in the international process. For those seeking further background information, a number of excellent books may serve as a more comprehensive guide to each of the negotiation processes: Litfin (1994) and Parson (2003) on the Montreal Protocol, Grubb et al. (1999), Oberthür and Ott (1999) and Newell (2000a) on the Kyoto Protocol, and Bail et al. (2002) and Andrée (2007) on the Cartagena Protocol.

The three cases studies below concentrate primarily on business actors from North America and Europe. The reason for this selective focus is partly a pragmatic one, to help manage the complexity of a research process that has covered multiple industries, countries and issue areas.

It also reflects the centrality of the United States (US) and the European Union (EU)[1] in ozone, climate and biosafety politics. Other countries have, of course, played an important role in the negotiations (e.g. Japan particularly so in the ozone and climate talks, and the Like-Minded Group of developing countries was the key *demandeur* for the biosafety treaty), but the issues at stake in the transatlantic conflict over how to protect the ozone layer, avert further global warming and ensure biosafety have been central to the international political dynamic in each of these cases.

The subsequent chapter (2) sets out the theoretical approach and analytical framework that guides the empirical analysis in this book. It reviews the existing debate on business power and introduces the neo-pluralist perspective that informs this investigation of international environmental politics. The three chapters of Part II provide detailed case studies of the international politics of ozone layer depletion (chapter 3), global climate change (chapter 4) and agricultural biotechnology (chapter 5). The final chapter (6) reviews the findings of each of the case studies and comments on the implications for the study of business power and business conflict in international politics.

2
Business Power and Business Conflict: A Neo-Pluralist Perspective

Corporations are said to 'rule the world' (Korten, 1995). As economic globalization reduces barriers to global investment and trade, international business is gaining ground. Some analysts of International Political Economy (IPE) have deduced from this that business is becoming ever more powerful vis-à-vis civil society and at the cost of state autonomy. At the height of the globalization debate in the 1990s, 'state retreat' (Strange, 1996) and 'power shift' (Mathews, 1997) were popular images that seemed to capture this new global reality (see also Barnet and Cavanagh, 1994; Falk, 1997; Rosecrance, 1999). Not all were convinced by this 'hyperglobalist' (Held et al., 1999) perspective, however. International Relations scholars rooted in the discipline's statist tradition have long argued that states ultimately control multinational corporations (Gilpin, 1975), and that globalization has done little to undermine state power (Waltz, 2000). Most recently, Daniel Drezner reformulated this argument in the context of global regulatory politics, claiming that 'the great powers … remain the primary actors writing the rules that regulate the global economy' (2007: 5).

Much of the recent debate in International Relations and IPE has oscillated between these opposing views. The trouble with this common framing of the globalization debate is that it forces discussions of business power and global governance into the straightjacket of a 'zero-sum game'. Accordingly, we are either witnessing a rise in the power of business and a decline in state authority, as the strong globalization thesis has it, or the opposite trend must be occurring, with states retaining control over

16

business and indeed the very process of globalization. This book is an attempt to depart from these dichotomous positions. What is needed is a better understanding of business power, its nature, its sources, as well as its limits, and how corporations interact with states and nonstate actors in the governance of the global economy. Business may play a more prominent international role in an era of globalization, but quite how business power plays out in international politics, and how it affects global governance is a less straightforward story than either hyperglobalists or globalization sceptics suggest.

While being situated in debates on globalization and global governance, this book takes a narrower focus: it examines business power in the field of international environmental politics. The fluctuating fortunes of environmental protection have been central to recent debates on how globalization affects our capacity to govern at the global level. Ecological crises such as acid rain, species extinction and climate change graphically illustrate the dangers of a globalized world of unfettered markets and weak governance. The last four decades have seen a flurry of international efforts to counter global ecological threats. A rising number of MEAs have been created, and several international organizations have taken on environmental roles. But whether these can provide effective global governance remains an open question.

What role does business play in international environmental politics? How powerful are business actors vis-à-vis states and nonstate actors, and to what extent, and how, do they influence outcomes in international policy-making? To answer these questions, three of the most high-profile issues in recent environmental politics – ozone layer depletion, global climate change and agricultural biotechnology – will be examined below (chapters 3, 4 and 5). To set the theoretical context and analytical framework for the case studies, the following section introduces the neo-pluralist perspective on business in international politics and develops it against the background of wider debates on the concept of power, its different dimensions and the role of agency and structure in International Political Economy. The neo-pluralist perspective argues that business is in a powerful, even privileged, position in international environmental politics, but does not have a dominant influence over outcomes. Countervailing forces limit its influence, particularly those that arise from divisions and conflict among business actors.

Neo-pluralism holds that business power is a contingent concept that needs to be studied in its historical, issue-specific, context. The second section below therefore places the discussion of business power and business conflict in the context of international environmental politics and discusses the factors that determine patterns of business power and business conflict in this policy area. To understand how business actors seek to influence international political outcomes and involve themselves in international processes, we need to develop a clearer understanding of how business actors form interests and develop corporate political strategies for international environmental politics (see third section, below).

A neo-pluralist perspective on business power

Power: concept and dimensions

Political analysis, in its general sense, is concerned with the political dimensions of any form of social interaction. Broadly defined, it is about 'the distribution, exercise and consequences of power' (Hay, 2002: 3). Power, therefore, is the central concept of political analysis, but remains deeply problematic in political science, as much as in International Relations and IPE. Little has changed since Robert Gilpin (1981: 13) observed over two decades ago that the concept of power is 'one of the most troublesome in the field of international relations'. Indeed, as Lukes (2005: 477) reminds us, power has long been considered to be an 'essentially contested' concept. Its meaning is intimately tied up with the fundamental assumptions that underpin our judgements about whether power is present in empirical situations, and how we should think about it. These 'value assumptions' inevitably provoke debate about how we should conceptualize power in political analysis.

Max Weber's classic definition of power as 'the probability that one actor in a social relationship will be in a position to carry out his will despite resistance, regardless of the basis on which this probability rests' (1964: 152) still serves as a widely used starting point in political analysis. In the behaviouralist tradition of political science, it has led to a focus on visible conflict between actors and specific outcomes in decision-making processes. But empirical research and theoretical debates have unearthed several problems with this widely used relational approach: conflict between different actors may not always be visible; power may exist without it being exercised in decision-making processes; and even where political interaction can be observed, establishing causal links between the exercise of power and policy outcomes is wrought with methodological problems (Dür and De Bièvre, 2007; Shapiro, 2006; Woll, 2007).

Alternative conceptions of power have stressed its structural dimension. Much of the power debate in International Political Economy has indeed focused on whether power should be thought of in relational or structural terms (Strange, 1988; Guzzini, 1993). More recently, sociological approaches have added a further dimension, in the form of discursive (Litfin, 1994) or constructivist (Guzzini, 2005) conceptions of power. Barnett and Duvall (2005) provide a different taxonomy, opting for four types of power: compulsory, institutional, structural and productive. What is common in all these recent reconceptualizations of power in international relations is the recognition that the different dimensions of power are not mutually exclusive categories. Although the main theoretical traditions attach greater importance to just one type of power, political analysis needs to proceed with an integrated view of power. As Barnett and Duvall argue, 'power works in various forms and has various expressions that cannot be captured by a single formulation' (2005: 41). The implication of this is clear: 'Scholars can and should draw from various conceptualizations of power that are associated with other theoretical schools' (ibid.: 45).

A similar sense of pragmatism has also returned to the study of interest group politics and power. When analysing the power of business actors, or any other nonstate actor, we also find that different dimensions of their power are simultaneously present, and that these forms of power need to be seen in their relevant context (Dür and De Bièvre, 2007). Just like states, business actors can be said to possess relational, structural and discursive forms of power (for a similar categorization, see Lindblom, 1977; Fuchs, 2005). The relational dimension of power is closest to Max Weber's definition in that it presumes a relational context in which business actors achieve their goals by pressuring state actors. Influence is, at least in principle, an observable phenomenon. It directs researchers to investigate policy outcomes and to establish causal links with input factors such as lobbying and political campaign funding. Underlying this perspective is the notion that power is based on the possession of certain resources (e.g. financial, informational) that can be brought to bear in the policy process. In that it focuses on measurable power resources and observable conflict, the relational concept of power comes closest to an empirical research programme on business power. But by ignoring invisible forms of power – for example, power that is present but not exercised, and power that prevents rather than influences decision-making – it tends to exclude important aspects of social reality.

It is this exclusionary tendency in relational concepts that has led to structural approaches to business power in IPE. At the heart of most structural approaches is the desire to bring out the hidden faces of power. Business is powerful not because it gets what it wants by lobbying state actors, but by setting the parameters of public policy. The power of business resides in the ability of firms to restrict policy-making and to determine the range of policies that are being fought over in the public domain. Bachrach and Baratz (1970) speak of 'non-decisions' in their famous study of agenda-setting power. Structural power is, therefore, 'invisible' as it is not being exercised in open competition for influence over policy-making. Instead, researchers need to make it visible by unmasking the underlying political-economic power structures that privilege business interests over others. Structural approaches have thus extended the analytic gaze, but by shifting from visible to invisible manifestations of power they have also complicated the empirical study of business influence. Links between power structures and non-decisions are difficult to establish. Moreover, the assumption that political agendas are limited by corporate power is in itself based on a questionable normative position of what ought to be on the agenda. Presuming the existence of what remains an elusive form of structural power can thus easily distort the assessment of how powerful business is. Nevertheless, we need to consider how business may exercise power indirectly, by constraining other actors and their capabilities.

Most structuralist approaches in IPE have tended to focus on material power structures. A more recent development has been to emphasize the discursive foundations of business power. Building on sociological theories of politics, discourse-based theories view business actors just like any other actor as being engaged in communicative processes that shape interests, identities and norms. Business actors are constantly seeking to legitimize their corporate interests and employ discursive strategies to shape public perceptions. Their authority rests not just on their command over economic resources but on their ability to shape the ideational foundations of society and the economy. Through communicative action, corporations seek legitimacy for their role as economic producers and political actors. Discursive power, in this sense, reinforces structural, and to some extent relational, power (Fuchs, 2005: 153–4). But as with structural power, the discursive dimension of corporate power is difficult to pin down in empirical research and remains an elusive category. If it serves to restrict agendas and constrain policy options, then discursive power will be difficult to trace empirically as it does not manifest itself in open, observable, conflict.

The discussion so far has suggested that business power needs to be viewed in its multiple dimensions. The challenge, of course, remains to operationalize a multi-dimensional concept of power. Recent debates on the concept of power, sparked not least by the publication of an extended version of Lukes' famous 1974 book *Power: A Radical View* (Lukes, 2005), reinforce the perception that we are far from resolving many of the underlying dilemmas that bedevil the study of power. But these conceptual and theoretical difficulties should not preclude us from engaging in context-specific empirical studies of business power in international politics. Indeed, as Ian Shapiro suggests, 'the condition of the power literature is such that most of the interesting social science questions are now best thought of as empirical' (2006: 146). By reflecting on the multi-dimensionality of power and its issue-specific characteristics within a given policy domain, we may hope to contribute a piece to the larger jigsaw puzzle that is the study of business in international politics.

Towards a neo-pluralist perspective on business power

This book advances a neo-pluralist understanding of business power. Before we can develop this perspective in the context of environmental politics, we need to briefly revisit the early debate on interest groups and business power, particularly between pluralists and their critics. The study of nonstate actors in international relations is, to a large extent, rooted in the study of interest groups in domestic politics. American politics in particular, with its vast range of organized interests and a relatively open process of interest competition in policy-making, has been a major testing ground for competing theoretical approaches. Much of the early political science literature on US interest groups arose within the tradition of the pluralist theory of the state (Key, 1942; Truman, 1951) and the economic interpretation of politics (Beard, 1934). It viewed the state primarily as a site of struggle for interest group influence, and state policies as reflections of the underlying balance of power between different interest groups. Although most pluralists did not concern themselves with international relations, some of their work, particularly on tariffs (Schattschneider, 1935), took on a more distinctive foreign policy dimension and thereby contributed, albeit indirectly, to a more overtly international theory of pluralist interest group politics (especially Beard, 1934; see Nowell, 1996: 192–3). Pluralism has been the target of much criticism in later political science, but it is still possible to derive important theoretical idioms and ideas from this tradition, as long as we reflect on its limitations and the special circumstances of international relations.

After the Second World War, with the arrival of behaviouralist approaches and quantitative methods in political science, pluralists developed more refined ways of studying the involvement of interest groups in policy-making (for an overview, see McFarland, 2004). Business interests featured in this work but did not always receive the attention they deserved. Pluralists tended to assume that business groups were just like any other interest group, with the exception perhaps that they commanded more financial resources in their lobbying efforts. In their view, ultimately, any group's attempt to dominate politics would result in the mobilization of countervailing interests and forces, thus contributing to the open nature of the political process. For pluralists, interest competition ensured the proper functioning of democracy.

Whether business just had more money or possessed other, structural, advantages became a hotly debated question from the 1960s onwards, and came to define one of the main differences between pluralists and their critics. Pluralists came under increasing attack as scholars identified various ways in which corporate interests 'captured' the policy process in advanced industrialized countries and created 'iron triangles' with legislatures and bureaucracies (McFarland, 2004: 32–5). In their view, certain business groups, based on their economic clout and contribution to economic welfare, were able to crowd out other interests and limit the state's autonomy. Miliband (1976) and Jessop (1982) arrived at similar conclusions based on a Marxist theory of the state, suggesting that elite links and capital's structural power allowed business to exercise a dominant influence in public policy. Power theorists such as Bachrach and Baratz (1970) and Lukes (1974) underpinned these arguments with a structural theory of power, according to which business is able to control political agendas, limit the range of policy options available to policy-makers and curtail public discourses that challenge the predominance of capitalist interests. In this view, democracy is a sham, at worst, or under threat, at best.

The debate of the 1970s clearly demonstrated the weaknesses of the original pluralist theory of interest groups. Its fundamental assumption that an open system of interest group competition would produce a near equilibrium of political forces and eliminate undue influence of particular interests ran counter to the growing realization that significant and systematic disparities in influence characterized politics in industrialized countries. Coming from the pluralist tradition, Charles Lindblom (1977) gave this argument a pointed expression with his notion of business as a 'privileged' interest group. Lindblom sought to account for such disparities by conceptualizing business as being in a separate category of interest groups. Business was privileged because it participated in normal pluralist politics *and* exercised structural power that limited the scope for normal politics. The key source of the business community's structural power lay in its ability to shift investment to countries with a more favourable business environment, thus forcing governments to compete with each other for capital.

Although it was only implicitly evident in Lindblom's work, the structuralist critique of pluralism forced a greater focus on the international dimensions of corporate and state power. If, as structuralists argue, capital has become more mobile and financial markets more integrated internationally, then globalization only serves to further accentuate the limits on state autonomy (Andrews, 1994). The pluralist notion of interest group competition would therefore be of even less value in the context of international economic integration. Indeed, the early debate on globalization produced a wealth of studies that supported the basic structuralist argument, namely that global economic integration strengthened business power at the expense of state autonomy (Ohmae, 1995; Stopford and Strange, 1991; Strange, 1996; Schmidt, 1995; Falk, 1997). But the effects of globalization turned out to be more complex than originally thought, and further studies produced evidence against a strong structuralist interpretation of global business power. In the end, the main arguments between pluralists and structuralists merely seemed to repeat themselves in the context of the globalization debate (Wilson, 2006: 40). The shift to the international level did not lead to new theoretical approaches but helped to clarify certain points of contention.

Critics of the strong globalization hypothesis have made three arguments that are of relevance in this context. First, while international economic integration is progressing, this trend has not necessarily locked states into a 'regulatory race to the bottom'. Despite the popularity of the idea that globalization leads to a downward spiral of deregulation, the evidence for it is, at best, patchy and unsystematic. Neither has globalization simply forced states to abandon policy autonomy in key areas such as welfare spending (Garrett, 1998; Glatzer and Rueschemeyer, 2005) nor have capital mobility and free trade led to deregulation in the environmental field or the creation of so-called 'pollution havens' (Jaffe et al., 1995; Neumayer, 2001). States in the industrialized world retain a high degree of policy autonomy, and they have sought to strengthen their ability to govern the global economy by creating new international governance mechanisms (Hirst and Thompson, 2000). Globalization thus should not be seen as forcing states and business actors into a 'win–lose' scenario, with power shifting between them in zero-sum fashion (Weiss, L., 1999).

Second, questions of power, governance and policy autonomy are best studied in the context of specific issue areas. As Theodore Lowi (1964) already argued in an influential review of one of the early systematic studies of American business influence, the pattern of interest group politics is shaped by the nature of the policy issue, and the nature of business involvement in politics varies from issue to issue area. This insight also applies to business engagement with international politics, and has, if anything, been reinforced by the globalization debate. Globalization itself is a highly uneven process, involving patterns of integration as well as fragmentation (Steger, 2003), and power and governance need to be contextualized within issue-specific domains.

Third, economic globalization has gone hand in hand with the expansion of political agency across boundaries. The rise of new types of transnationally organized social actors, a phenomenon often referred to as transnational or global civil society, has opened up international policy-making and created new opportunities for political contestation outside the states system (Keck and Sikkink, 1998; Florini, 2000; O'Brien et al., 2000). Some of these actors have directly challenged the power and authority of business actors (Wapner, 1996), leading to new forms of civil regulation of multinationals by consumer groups and environmental activists (Murphy and Bendell, 1997; Newell, 2000b). Interaction between transnational NGOs and MNCs has in some cases led to the creation of private forms of global governance (Cutler et al., 1999; Falkner, 2003). This new pattern of transnational interaction, competition and alliance formation has rendered global politics 'more fluid and open' (Cerny, 2003: 156).

Insights from the recent globalization debate therefore suggest that a more pragmatic approach is needed to the study of business power. It would recognize that power can take on relational and structural dimensions, but would be cognisant of the methodological difficulties of researching the 'second' and 'third face' of power (see McFarland, 2004: chapter 8). As I argue in this book, a neo-pluralist perspective on business power provides such a pragmatic approach. It views power as multi-dimensional, taking on both visible and invisible forms, but argues that power needs to be contextualized and studied in specific policy domains. Business may be in a privileged position compared to other nonstate actors, mainly due to its economic power, but this does not in itself allow it to determine outcomes in international processes. Neo-pluralists argue that business power is constrained by external and internal forces: countervailing forces outside the business sector (e.g. civil society) and tensions and conflict within the business sector. In accepting the structural dimensions of power, neo-pluralism departs from earlier pluralist conceptions. But in insisting that questions of power and influence need to be resolved through empirical study, and that structural power needs to be translated into the policy process through the agency of business actors, it rejects the often far-reaching but methodologically questionable claims by structuralists.

To understand why business power is limited in this way, and why international political processes should be presumed to be open-ended, we need to briefly consider the countervailing forces that prevent business influence from becoming dominant. They can be found in the resilience of state power and the proliferation of new transnational political actors, but most importantly in the heterogeneity of the business sector itself. Neo-pluralism's key insight in the international context is that the diversity of business interests, combined with the persistence of business conflict, serves to limit business power overall.

Countervailing forces can be found outside and within the business world. With regard to the former, states retain their status as loci of authority not only in core state functions such as security, as structuralists such as Strange (1988) have acknowledged, but also remain powerful gate-keepers and providers in other policy areas that are more open to the influence of nonstate actors (Drezner, 2007). Furthermore, new transnational actors have come to challenge the legitimacy and authority of business actors even in domains where they can be said to be in a privileged position. New channels of transnational communication and campaigning have empowered social actors, even though they often rely on only limited financial resources and lack access to established policy networks (Tarrow, 2005). Particularly in the environmental field, grass-roots and transnational campaigns by activist groups have undermined the legitimacy of multinational firms and induced change in corporate behaviour (Wapner, 1996). To be sure, interest group competition in transnational and international realms is rarely conducted on a level playing field. Global political space is not entirely pluralistic, but existing balances of power between different transnational actors vary across different policy domains and are more fluid and unstable in an era of globalization, leading to a more open-ended process of global politics.

The pluralist message is further reinforced when we consider dissent and conflict within the business sector. The straightforward but important insight that neo-pluralism offers is that business is often divided on matters of international policy and corporate strategy, and that business should therefore not be treated as a monolithic bloc. The corporate sector may of course, in some vague sense, represent a capitalist class interest, but this claim amounts to little more than a truism that is of limited analytical value in the empirical study of business influence in specific policy contexts. Indeed, if we want to understand the sources and limits of business power and influence, we need to disaggregate the business sector and analyse its constituent parts, down to the level of the firm. For particular business interests to exercise a dominant influence, achieving business unity is an important but highly demanding condition. Business conflict thus serves as an important brake on business influence in international politics.

This line of thinking has become known in IPE as the 'business school model' (Skidmore, 1995; Skidmore-Hess, 1996). Societal approaches that focus on the domestic origins of foreign policy have been at the forefront of this development. By identifying cleavages that exist within the business sector, Frieden (1988) and Milner (1988) have explained the shifting patterns of business support for free trade and protectionist policies in the US and elsewhere. Rogowski (1989) uses factor endowments theory to analyse how the gains and losses from international trade are distributed between different economic sectors, and how those distributional effects influence business preferences in trade policy. Others also focus on business factionalism in their analyses of US foreign policy, with domestic and internationalist coalitions vying for influence over state policy (Gibbs, 1991; Cox, 1994; see also the contributions in Cox, 1996).

The main focus of the business school model has been to explain outcomes in foreign policy and international politics from the bottom up. By reversing the perspective, we can also capture the ambiguous effect that globalization has had on business actors. While international business has been the main beneficiary of ever greater economic integration, it has also become more exposed to new political demands and pressures that globalized politics has created. The nature of the international political process has changed due to globalization, resulting in a more open and fluid process of policy-making involving an ever greater number and diversity of actors. Whether it is the international politics of trade and finance or new issue areas such as blood diamonds or genetically modified food, business actors are now faced with a large number of civil society actors that seek to create new international norms and affect corporate behaviour directly by challenging the power and legitimacy of business (Vernon, 1998; Bomann-Larsen and Wiggen, 2004). The advent of new information technologies has significantly reduced the costs of 'presence' and 'voice' in global politics, and transnational campaign groups have skilfully leveraged their social and discursive power through the use of symbolic politics. As political globalization progresses, established positions of power and influence are being challenged and redefined. This, as Cerny points out, reaffirms the neo-pluralist insight that 'those social, economic and political actors with the greatest access to material and social resources generally marshal those resources in uneven and complex ways in order to pursue their own interests as effectively as possible in what is still a relatively open political process. They predominate, but they do not necessarily control' (2003: 156).

It should be noted that to place business conflict at the heart of the neo-pluralist perspective does not mean that such conflict is assumed to be the predominant pattern of behaviour among firms. Indeed, business actors routinely seek to limit the potential for conflict and competition in an effort to stabilize the organizational field in which they operate. Students of business organization have long argued that the desire to reduce price competition and stabilize organizational fields is central to the strategy particularly of large multinational enterprises (Fligstein, 1990; Spar, 2001). Likewise, business actors will seek to minimize differences and tensions between them in their efforts to shape international political outcomes. On issues that affect most corporations in an equal way, business unity will be easier to achieve. But on other issues that have differential effects on individual firms – and regulatory politics is one such area – the potential for disunity and conflict can never be excluded. It is therefore analytically preferable to treat the question of business unity or conflict as an empirical question, not as a given.

We will return to the question of business unit and conflict again at the end of this chapter, but now turn to the specific circumstances of environmental politics at the international level.

Business power and conflict in international environmental politics

Dimensions of business power

In what ways can business be said to be powerful in international environmental politics? Our analysis relies on a multi-dimensional concept of power that is contextualized in an issue-specific and historical setting. We therefore need to consider the specific characteristics of international environmental politics, in order to place the discussion of business power and business conflict in its proper context.

The *relational concept of power* focuses on the power resources that actors employ in order to achieve the political outcomes that they seek, if necessary against the wishes of others. Power as a relational concept assumes that different actors compete with each other over influence, that they employ resources that can be measured, and that at least in principle we should be able to observe the exercise of power in situations of open conflict. In international politics, lobbying state actors is the standard way for nonstate actors to influence outcomes. To do this, they employ a variety of resources that they have at their disposal: financial, organizational and institutional.

One of the main strengths of business actors is the often unrivalled financial resources that they possess to fund lobbying activities, particularly when compared to the funds available to NGOs and other nonstate actors. Financial resources allow business interests to be represented in a variety of international contexts and over a long period of time. Given that there are now over 200 international environmental treaties in existence which hold regular meetings of parties or subsidiary bodies, sufficient financial resources are of critical importance to those who wish to exercise a more systematic influence on such forums. Larger MNCs, particularly, can afford to employ professional lobbyists or buy in ad hoc legal or scientific expertise that allows them to follow and influence international negotiations. Business organizations have spent multi-million dollar budgets on advertising campaigns to inform the wider public about their positions on environmental matters and to influence policy-makers on critical issues in international negotiations, as happened in the US when the Global Climate Coalition (GCC) spent $13 million on a public relations campaign targeted against the Kyoto Protocol on climate change (Levy, 2005: 83).

The business sector's relational power also includes its organizational strength in engaging with international environmental negotiations, allowing a diverse range of business actors to bundle their resources and develop more targeted lobbying campaigns. Often, individual companies cannot afford to sustain lobbying efforts of their own or are unable to follow all ongoing international processes. One of the key strengths of the business sector has therefore been the capacity to unite large industry segments under larger umbrella organizations or create issue-specific lobbying organizations that address controversial environmental issues. As one of the largest international business associations, the International Chamber of Commerce has closely followed many international environmental negotiations and sustains a regular lobbying and communication team to engage with new and upcoming issues. Sectoral or regional business associations, such as North America's Chemical Manufacturers Association (CMA) and the Union of Industrial and Employers' Confederations of Europe (UNICE), have also represented business interests in international forums. More recently, and especially since the 1992 Rio 'Earth Summit', a new generation of business associations has emerged to provide a more focused form of representation in environmental affairs, including the WBCSD, the Business Environmental Leadership Council (BELC) and the GCC.

In seeking to shape international outcomes, business actors can also build on their established positions within domestic policy networks that provide them with often privileged access to decision-makers. Business representatives routinely interact with governmental representatives of regulatory agencies and are in close contact with ministries dealing with economic affairs, and can rely on these contacts to seek to influence governmental positions in international negotiations. In most industrialized countries, business leaders have been able to rely on close working relationships with representatives of business-friendly government ministries, such as those for industry, agriculture, commerce or trade, or science and research. In this way, business groups' privileged position at the domestic level can translate into a privileged form of access to governmental actors at the international level, providing business actors with information about ongoing negotiations and support within national delegations.

When compared to other actors, business actors have considerable, and often unrivalled, power resources at their disposal. However, whether these allow business to prevail over other nonstate actors in international environmental politics is far from certain. In fact, experience with recent MEA negotiations suggests that environmental NGOs have on many occasions been a good match for the business lobby. NGOs' ability to mobilize mass support and claim public legitimacy can compensate for their lack of financial resources. NGOs have developed transnationally coordinated campaigns to overcome some of their resource constraints and are able to use new and inexpensive forms of electronic communication to maximize their impact at the international level. Moreover, business links to industry and trade ministries are of only limited value when international negotiations are led by environmental ministry officials, as is routinely the case. In the negotiations on the Cartagena Protocol on Biosafety, for example, business representatives initially felt ill-represented in a process that was dominated by scientists and regulatory experts representing mostly environmental ministries (see chapter 5). The peculiar nature of international environmental politics has thus served to reduce the impact of many of the power resources that business actors routinely employ, thus helping to create a more open and pluralistic field of interest group lobbying.

Business is said to possess *structural power* because of its central role in the economy, as the main source of economic growth, employment and innovation. Policy-makers are constrained in their choice of policies and policy instruments by the need to ensure that business regulation does not put an undue burden on the business sector. If policy-makers ignore the needs of business, firms will respond by reducing investment or relocating to other countries, thus harming the legitimacy of the state and the electoral chances of politicians. This classic structuralist argument can also be applied to environmental politics. Just as in any other form of regulatory politics, environmental policy-makers will be constrained by the knowledge that overly strict environmental standards or rules will harm core economic interests, create a competitive disadvantage and encourage businesses to move abroad. In designing political responses to environmental crises such as climate change, decision-makers will therefore seek to either gain the support of business leaders or choose instruments that are considered business- or market-friendly and that give sufficient flexibility to the business sector to adapt in the most cost-effective manner. In extreme cases, the structural power of business will deter policy-makers from taking environmental action altogether. Thus, structural business power limits the ability of states to impose solutions that may be environmentally desirable but threaten to violate the fundamental interests of business. Structural power has a disciplining effect on politics, by keeping certain environmental issues and solutions off the political agenda.

A more specific form of structural power exists in environmental politics due to the central role that technology and technological innovation play in dealing with environmental issues. Companies can be said to possess 'technological power' (Falkner, 2005) because of their ability to produce technological innovations that help solve environmental problems. Because in market economies, business plays a central role in determining the level of investment in research and development and directs technological innovation and diffusion, corporate decisions on such investment take on an imminent political role. Policy-makers may seek to use regulatory instruments to encourage technological innovation in an effort to reduce harmful emissions or change industrial processes. However, in designing such regulations, they depend on technological knowledge – about the availability of technological alternatives, the lead times for introducing new technologies and the associated costs of product or process change. This knowledge base is largely controlled by companies themselves, and they are therefore in a privileged position when it comes to shaping perceptions of the economic costs and technological barriers in solving environmental problems caused by industrial processes. Technological power enables business to shape regulatory discourses, particularly when it comes to the design and phasing of regulations.

How important is the structural power of business in environmental politics? Does it allow business to prevent international outcomes that are harmful to its core interests? There is no question that the central importance of modern industrial processes to the functioning of the global economy puts certain limits on how far international environmental regulation can go. Drastic and immediate efforts to eliminate greenhouse gas emission from fossil fuel-based sources will threaten not just the core interests of the oil and coal industry but also of major transport and manufacturing industries that rely on cheap and steady energy supplies. Indeed, many international environmental treaties have failed to demand timely action against environmental threats, suggesting that the interests of business have trumped the needs of the global environment. But the problem with arguments that rely on such structuralist reasoning is that they tell us little about the specific dynamics of international environmental policy-making and make it difficult to disentangle business power from other sources of political inertia. Structural power is by and large invisible, and if it is to become a meaningful category in our analysis, we need to identify the processes through which it is brought to bear on policy-makers in the international process.

To say, therefore, that the root cause for the weakness of many international environmental regimes lies in structural business power serves to shortcut the analysis and to ignore other, and potentially more important, causal factors. Weak environmental regimes also reflect the inherent difficulties of multilateral policy-making involving often up to 190 sovereign states, as well as the complex political trade-offs that societies have to make between environmental protection, technological innovation, economic development and poverty reduction. Furthermore, a number of environmental regimes have been created often against fierce business opposition, suggesting that even if regime-building processes are incomplete, they may nevertheless hurt business interests. Indeed, the very fact that an international environmental agenda has been created over the last four decades, and in many instances against business interests, suggests that structural power has not allowed business to control the political agenda as such.

Discursive power is the third dimension of power that is of relevance in international environmental politics. Following a sociologically inspired reasoning, we can conceive of actors including firms as engaging in communicative practices – also referred to as discourses – that revolve around ideas, identities and values. In trying to shape social understandings of the problems to be addressed and the ideas and norms that should guide policy-making, actors employ discursive strategies and can be said to possess discursive power. In the environmental field, discursive forms of power are closely tied to the special knowledge that is involved in identifying ecological problems and defining technological solutions (Litfin, 1994). Actors engage in discursive practices when they debate and define the boundaries of what is relevant knowledge, including the policy options available to decision-makers. Viewed in this sense, business has discursive power due to its central position in the technological innovation process. Discursive power becomes akin to technological power as defined above, as it sets the parameters of what is perceived as a technologically and economically feasible solution. Regulatory discourses are thus strongly influenced by the discursive power of firms rooted in their privileged economic position. Business actors also seek to shape public discourses on environmental issues by engaging in more direct forms of communication, through advertising and public relations campaigns. Major global firms now routinely communicate to the public their specific approach to environmental issues, be it on global issues such as climate change or product-specific aspects of their business operations. Commercials, corporate publications and annual reports, as well as appearances by corporate leaders in parliamentary hearings and public debates, all serve the purpose of injecting a business perspective into the environmental debate, in order to shift the regulatory discourse in the desired direction.

That business groups have sought to expand their discursive power as global environmental threats have increasingly undermined their public legitimacy is now widely acknowledged (Fuchs, 2005: chapter 7). But measuring the effect of discursive power is wrought with similar methodological problems as in the case of structural power. The legitimacy of business groups remains constantly threatened by counter-discourses that have sprung up in reaction to economic globalization (Prakash, 2002). And although efforts to shift environmental discourses in a more business-friendly direction have borne some fruit, as can be seen in debates on sustainable development (Schmidheiny and BCSD, 1992; Eden, 1994), global political debates and regulatory discourses have shown themselves to be too fluid to be controlled by corporate actors.

Business conflict in international environmental politics

Business conflict arises in international environmental politics because of the differential effects that international regulatory measures have on individual companies or industries (Falkner, 2001). Environmental regulations can take on many different forms and include a variety of mechanisms, including process and product standards, international monitoring or certification schemes, identification and documentation requirements for international trade and information exchange, targets and timetables for the reduction or elimination of harmful emissions, and emission trading schemes, among others. What they all share in common is that they rarely have a uniform effect on business as a whole, but target specific groups of corporations or industrial sectors, create new markets or adjust existing ones. The aim of regulations is to change corporate behaviour in a specific and targeted way, and it is this that creates uneven effects on business overall, potentially leading to a divergence of business interests, and even conflict. Business actors can therefore be expected to form interests and political strategies on international environmental politics that seek to limit the costs of regulation or maximize its benefits.

Several forms of business conflict can be identified with regard to international regulation, norm-setting and regime building. First, as suggested by studies on international trade policy (Frieden, 1988; Milner, 1988), a basic dividing line exists between international and national firms. International firms are more likely to support international rule-setting and the harmonization of national regulations. The latter have traditionally favoured protectionism in trade policy and are more likely to oppose international rule-setting in environmental affairs. Firms that operate in different national markets and depend on the unhindered flow of goods will place a higher value on creating a level playing field than those that are concerned primarily with national markets and competition from abroad. This does not mean that international firms will always support international environmental regulation. They are likely to do so only where it provides them with a competitive advantage, by reducing the transaction costs of operating in multiple regulatory environments, and by raising the regulatory costs of competitor firms that operate in countries with lower environmental standards. This divide can be seen in the politics of ozone layer protection, where the highly globalized chemical industry was the first sector to support international restrictions on ozone-depleting substances, while many domestic industries that used these substances remained opposed to international restrictions for much longer (see chapter 3). Vogel (1995) has referred to this effect as 'trading up', where international firms promote the adoption of higher environmental standards in an effort to create a global or regional level playing field.

A second, and closely related, form of business conflict can arise between technological leaders and laggards in the same industry or economic sector, be it nationally or internationally organized. In this case, the dividing line is found between competitors in a given market segment that are likely to experience differential effects of regulation due to their uneven ability to comply with new standards. If market leaders can hope to lower their compliance costs relative to their competitors, then an increase in regulatory standards and compliance costs may shift the competitive balance in their favour, thus making regulation more acceptable to them. The degree to which companies can respond to new environmental regulations through technological innovation will thus be an important factor in determining their overall political strategy. In some cases, regulation can produce new markets based on technological innovation that would otherwise not have been commercially viable, and technological leaders can therefore use regulatory politics to create new business models and achieve competitive advantage (Porter and van der Linde, 1995).

A third form of business conflict can arise between companies that operate in different economic sectors but are linked together in so-called supply chains or production chains (more on this concept below). Wherever regulations target specific products or processes involved in their production, they will affect not only the sector that is directly responsible for the production or sale of the product but all companies that contribute to it at different points in the production chain that links suppliers of input factors, producers and retailers together. The important point to note is that regulation is likely to have differential effects on the companies that operate along the production chain, leading to divisions and competition between them. While companies operating at the consumer end of the chain (e.g. retailers) may support higher regulatory standards as part of their strategy to maintain consumer confidence or enhance their reputation, companies providing raw material inputs or intermediary products further down the chain may end up facing higher production costs without gaining any reputational benefits. For example, supermarkets in Europe and North America have generally supported higher food and environmental safety standards in food production, but smaller producers particularly in developing countries have experienced difficulties in meeting those standards in a cost-effective manner. European supermarkets were the first to ban genetically modified food from their shelves, against the wishes of biotechnology firms and agricultural exporters in North America (see chapter 5).

In sum, business conflict is an important feature of business involvement in international environmental politics. Whether it exists in reality or is only a latent threat to business unity depends on the nature of regulatory policies under consideration and the strategies that different companies form. For business conflict to become politically significant, business actors need to be able to identify the differential effects of regulations and integrate these perceptions into coherent political strategies. We therefore need to consider in more detail how companies form preferences and devise strategies, which is the subject of the next section.

From corporate preferences to corporate political strategy

Corporate preferences: economic and institutional dimensions

Identifying the preferences that corporations form with regard to international environmental policy is a first and necessary step in the analysis of business power and business conflict. It is essential for a better understanding of the conditions under which business actors develop political strategies and involve themselves in international politics, and when business conflict comes to characterize business involvement.

Recent advances in IPE have produced two major strands of thinking on business interests. The first deduces preferences from the competitive position that companies or industries find themselves in within global markets, and from the effects that shifts in international policy have on their position. The second takes the institutional environment into account that shapes corporate preference formation, and that acts as a filter through which economic interests are perceived. Both perspectives need to be seen together to arrive at a fuller picture of how international environmental politics affects corporate interests.

In the first perspective, corporate interests are defined purely in economic terms. They reflect the position of specific firms and industries within the wider competitive field that they are operating in, and involve a calculation of how different forms of international regulation affect their position and the market environment. Corporate preferences are thus assumed to be structurally determined, and they will translate into political lobbying strategies that seek to influence foreign policy accordingly (e.g. Frieden, 1988; Rogowski, 1989). Applied to the environmental field, corporate actors evaluate the impact that regulatory proposals have on their business and the competitive balance in the market. They will consider the extent to which they can adapt to new environmental standards through technological innovation, product changes and process modifications, and how their competitors will be affected. As discussed above, divisions in corporate preferences may emerge, between national and international firms, between technological leaders and laggards, and between different firms along the product chain.

The economic interpretation of corporate preferences is an important starting point in the empirical analysis, but is limited in significant ways. As institutionalist and sociological theories of the firm have argued, corporate interests are not simply a given that can be deduced from the competitive position of firms within markets. Instead, we need to view corporations as being embedded in institutional and social networks that give rise to shared understandings and values (Callon, 1998; DiMaggio and Powell, 1991; Fligstein, 2002). Corporations are themselves social institutions, with corporate leaders pursuing their own interests and values within the organizational and social context of the firm. They are also part of a wider organizational environment that informs and shapes perceptions of what preferences corporations should form and what is to be considered legitimate business interest. In this view, corporate preferences are dependent on frameworks of understanding that interpret economic factors but are themselves rooted in social and institutional contexts. Thus, even if market pressures and incentives are the same, companies may form different interests and adopt different strategies due to 'divergent understandings prevalent in the particular economic, political, and socioideological networks in which individual firm managers are embedded' (Pulver, 2007: 4).

A number of institutionalist and sociological arguments have been advanced recently to this effect (for an overview, see Smelser and Swedberg, 2005). Research on the variance in the organizational and strategic divergence among MNCs has highlighted their national embeddedness as a key factor that counteracts the convergence effect of economic globalization. The home-country of a multinational provides distinctive political, economic and cultural contexts within which its organizational structure and operational strategy evolve (Doremus et al., 1998). When expanding internationally, multinationals often export market and non-market strategies from their home-country context, including in the field of environmental management (Garcia-Johnson, 2000). And in dealing with national and international environmental politics, established patterns of business–government relations shape the nature of business lobbying. Several authors have noted that the corporatist political economy of Europe has given rise to more conciliatory and cooperative approaches by European firms in contrast to the more antagonistic political strategies of US firms (Hillman and Keim, 1995; Raustiala, 1997).

These institutionalist perspectives direct our attention to the cognitive processes that make up preference formation and the cognitive frames that inform actors' perceptions. No market structure can be grasped without some framework of understanding that gives meaning to economic facts, and economic reality does not present itself to corporate actors as a given. To borrow Alexander Wendt's famous phrase (1992), sociologists would posit that markets are what firms make of them. But a sociology of the firm that ignores economic structures risks missing the important connections that exist between markets, political environment and social context. It is in the interaction between them that corporate preferences are formed. As Levy and Newell (2000: 10) remind us,

> Business strategies are primarily driven by perceptions of economic interests, filtered through particular national lenses and constrained by specific political and social contexts that vary by issue over time. Economic, political, and cultural forces interact in complex ways to produce the outcomes on each of the issues.

The analysis of interest formation thus needs to be sensitive to both economic and institutional dimensions. Economic incentives and pressures are major drivers of business responses to emerging global political issues, but the social, organizational and political environment in which firms operate define how they perceive of their interest and define strategies.

Production chain analysis

The organizational fields that matter in our analysis are not only those that connect companies operating in the same sector (e.g. chemical industry, automobile manufacturers) but also those that combine companies from different sectors that cooperate in the production and distribution of specific goods. Such fields are known in political economy and organization theory as production chains or production networks. The concept first emerged in business studies, where authors such as Michael Porter (1985) focused on so-called 'value chains' to identify the stages at which corporations add value to economic processes that are connected across boundaries. Gereffi (1994) introduced the notion of the 'commodity chain' to identify new forms of transnational industrial organization that link households, corporations and states together within the global economy. Unlike Porter, Gereffi's work specifically focuses on structures of governance and power relations within commodity chains, and thus serves as a key inspiration for the political-economic study of business power and economic development. More recent work by Henderson et al. (2002) has developed the more ambitious concept of global production networks (see also Dicken, 2003). While it builds on Gereffi's work, it seeks to capture the more complex and horizontal forms of transnational economic linkages that evade the more vertical orientation of 'chain' approaches.

What all these concepts have in common is that they seek to identify transnational organizational structures that integrate different steps in the production process, involving different corporations and industrial sectors that operate independently from each other. Production chains or networks serve as a heuristic device to identify relations of power and control, the organizational environment of firms, as well as potential cleavages between different corporate actors in relations to matters of global concern. Production chains therefore provide a useful starting point in the empirical study of interest formation, as they help identify the relevant economic network and organizational field that is likely to be affected by proposals for international environmental regulation.

Production chain analysis advances three key insights that are of importance to the study of business power and conflict. First, in an age of economic globalization, production networks are commonly organized transnationally, involving producers, traders and retailers that operate across boundaries and in different national locales. The concept of the chain or network signifies that different corporate activities are functionally integrated so as to form a continuous process of production that binds different firms, sectors, and ultimately economies, together. Such chains can be found in a wide range of economic sectors, from automobiles, aircraft and computers to textiles and consumer electronics (see Dicken, 2003). They are particularly prevalent in environmentally sensitive industries, such as mining, energy and chemicals. Environmental regulations that target one sector or type of economic activity are likely to reverberate through such chains and draw in a much wider range of actors than originally targeted.

Second, production chain analysis points out that production chains and networks are based on relations of power and control between corporate actors. The globalization of production has led to an increasing dispersal of economic activities across different markets, which in turn has increased the need to organize production chains and control the corporate actors contained within them. Creating vertically integrated multinational corporations has been one answer to the governance challenge of globalization (Jones, 2005). Where such integration has not occurred, alternative forms of control have emerged, in the form of supply chain management and codes of conduct by powerful firms operating at different levels in the production chain (Cutler et al., 1999). To understand, therefore, where business power resides within globally integrated markets, production network analysis offers a useful analytical tool for mapping relationships of control in the global economy. In his work on commodity chains, for example, Gereffi (1994) distinguishes between two predominant modes of supply-chain control: *producer-driven networks*, in which multinational corporations centrally control the production system across the entire production chain, as is the case in technology-intensive industries (e.g. Airbus, Siemens); and *buyer-driven networks*, where large retailers and brand-based trading companies play a key role in managing decentralized corporate networks in different national markets (e.g. Wal-Mart, Nike). While this distinction does not fully match the often complex reality of international economic networks, it helps to identify key structural features of the global economy. In this way, production network analysis highlights the international integration of modern industrial production, points to structures of power within the corporate sector, and identifies potential sources of tension and conflict between different business sectors and firms.

Third, as David Levy (forthcoming) has argued, production networks are organizational fields in which processes of contestation and stabilization coalesce. They are situated in a wider social context, with governments, consumers, civil society groups and other stakeholders constantly reshaping the normative environment in which production takes place. At times, existing structures of corporate power and control are being challenged from the outside, and processes of corporate response, accommodation and restructuring lead to new arrangements that seek to stabilize production networks. We should therefore view networks and chains in the global economy as constantly at risk of contestation, with pressure from social and political actors as well as economic changes producing change in the ordering of corporate relations. Social contestation has grown in recent years, and a number of global production networks have been the subject of intense pressure over their human rights, labour and environmental standards. Such contestation has focused on, for example, sweatshops in textile production, fair pricing in the coffee industry, and the impact of oil companies on local communities (Tulder and Kolk, 2001; Kolk, 2005; Wheeler et al., 2002). A broader effort to redefine the terms of legitimate business practice in a global economy has been made in the context of debates on corporate social responsibility (Vogel, 2005).

In sum, production chain analysis provides a useful analytical device that allows us to assess how environmental regulations will affect a variety of business interests operating in different sectors; helps map the market structures and organizational environment in which business actors form their preferences and devise political strategies; and identifies important cleavages between different corporate and sectoral interests in the global economy.

From corporate preferences to political strategy

The analysis of preference formation is a necessary first step in the study of business power and conflict, but we need to go further and consider how business actors move from preference formation to political strategy and involvement in international processes. Preferences do not necessarily translate into political action, and even when they do, the translation process is not always straightforward. For business interests to become politically relevant, business actors need to form individual and collective political strategies, a process that involves aggregation of interests, choice of lobbying strategy and coalition building. Along this path, several complicating factors intervene, affecting the way in which business interests are injected into international politics. Not all interests may be fully represented in the end, some interests may prevail over others, and the choice of strategy will reflect not only the underlying preference structure but also the wider organizational and political environment in which corporations operate. To understand the role of business in international environmental politics, therefore, we need to consider the factors that shape the formation of corporate political strategies.

The idea that corporations might pursue political strategies has gained in popularity more recently. One of the limiting factors in the conventional business literature has been the tendency to draw a sharp distinction between market and non-market activities. But as several authors have argued (Baron, 2006; Levy and Newell, 2005), this distinction masks the reality especially among international businesses that operate in different political and legal contexts, and for whom successful political strategy is essential to their long-term success. For multinationals, but potentially also for national firms, political strategy becomes an integral part of doing business wherever policy-making can shape markets and change their competitive dynamic. Whereas traditionally, economics and business studies viewed the firm as an economic organization for which the political environment was externally given, more recent work has integrated the political role of firms and focused more on corporate attempts to shape the political environment out of economic objectives (Boddewyn and Brewer, 1994).

But pursuing corporate political strategies is costly. It requires strategic vision, an understanding of complex international processes, and the allocation of resources to follow and influence international politics over a sustained period of time. Not all firms have the resources to do this or can afford to spend them in this way. Only some firms will, therefore, develop international political strategies in the first place. For some, especially small and medium-sized companies, the cost of representation at the international political level and lobbying in international negotiations can be prohibitive. Regulatory politics at the international level can be a long, drawn-out and unpredictable process. The 1982 UN Convention on the Law of the Sea took over a decade to negotiate, and many multilateral environmental negotiations drag on for several years, in some cases not even leading to a tangible result. Faced with the open-ended and uncertain nature of international diplomacy, many firms will therefore simply choose not to be involved in these processes, or they will either join large business associations (e.g. the ICC) that offer to represent a wide range of business interests in multiple forums. Indeed, it is often only large, multinational corporations that can afford to sustain a lobbying presence in international contexts. MNCs are more likely to form independent strategies as they operate internationally and have more immediate interest in shaping international regulations that affect their business operations. For most other firms, especially small and medium-sized firms, the best and most cost-effective way to seek to influence the international political environment is at the national level, when international agreements are being implemented and translated into national legislation and regulations. Their focus will be on ensuring that such international rules, once they reach down into the domestic realm, do not pose an undue burden on business.

To be effective in international regulatory politics, individual business interests need to be aggregated and corporate political strategies need to be coordinated. This inevitably leads to collective action problems that are not that different from the problems faced in domestic contexts of group politics. To achieve interest aggregation and coordination in environmental policy, already existing or newly created business associations have taken on the role of international lobbyists on behalf of often wide-ranging business interests. Associations such as the ICC seek to represent business interests across industries and countries and in multiple international forums, while others such as the GCC have been formed to promote the interests of a more select group of companies on a specific policy issue. The former can claim to speak for business overall, but are often held back in their lobbying role by the need to pursue lowest common denominator positions given the wide range of often conflicting interests they speak for. The latter are able to create a more homogeneous and coherent lobbying strategy but can do so only on behalf of a more select business community. Even so, issue-specific business groupings are not immune to the effects of business disunity. The GCC, for example, which played an influential role in the early phase of international climate politics, was unable to prevent a major split among the leading oil companies in the mid-1990s and was disbanded in 2002 (see chapter 4).

The political strategies of firms are also affected by the political and institutional environment that they operate in. Business associations provide one, though not the only, institutional context. As economic sociologists argue, firms are embedded in a network of organizations, or an organizational field, that has a significant influence on how businesses perceive of their interest and define their political strategy (Fligstein, 1990). Comparative political scientists emphasize the importance of regional or national systems of political economy that structure relations between business and government, and thus lead to different types of corporate political strategy (Hillman and Hitt, 1999). In corporatist systems, comprehensive business associations enjoy a high degree of centralization and a privileged position in representing employers' interests vis-à-vis governments and labour organizations. They are thus in a better position than comparable associations in pluralist systems to organize and harmonize divergent business interests and can afford to commit their members to long-term policy objectives (Katzenstein, 1985; Streeck, 1992). Others have focused on subnational and transnational policy networks that exist in specific issue areas (Smith, 1993). In such networks, relatively stable groups of public and private actors repeatedly interact to develop public policy. Long-term membership in such networks may develop institutional characteristics and can become a dominant factor in shaping corporate political strategy.

In sum, when examining the role of business in international environmental politics, we need to consider not only the specific interests of business actors but also how these are translated into political strategies and lobbying efforts at the international level. All this takes place within a wider organizational field that is made up of the business, social and political environment within which firms define and pursue their interests.

The question of business unity and conflict revisited

As outlined in this chapter, the neo-pluralist perspective argues that business is in a powerful, even privileged, position in international environmental politics, but does not have a dominant influence over outcomes. It sees countervailing forces as limiting the influence of business overall, particularly as a result of business divisions and conflict arising within the business sector. The discussion so far has shown how environmental regulations create differential effects on individual business actors, and how even under conditions of similar external constraints different actors will arrive at different interests and strategies depending on the organizational environment within which they operate. Business conflict can thus be assumed to be a major characteristic of business involvement in international environmental politics.

Still, it can be argued that what we have considered so far are special cases in which corporate perspectives diverge, and that there are more fundamental business interests that all corporations share, irrespective of the distributive effects of regulatory politics. Competitive pressures and conflicting business agendas may be at work, but at a deeper level, all business representatives will share a common interest in maintaining an economic system that is relatively free of regulatory shackles. Therefore, business conflict need not be the predominant pattern of business involvement in international politics.

This argument takes us back to the opening discussion about neo-pluralism's departure from both statist and structuralist theory. While the former is on the whole not concerned with questions of business interests and conflict, the latter tends to replace the specific interests of individual business actors with the larger and analytically prior concept of class or capitalist interest. The main difficulty for structuralism has been to explain how an overarching and ultimately uniform business interest can emerge from the empirical diversity of business voices. The details of the longstanding debate on how this is achieved in historical contexts – whether through elite consensus mechanisms or the mediating role of the state – need not concern us here (for an overview, see Dunleavy and O'Leary, 1987), but it is important to note that one of the conclusions from this debate has been a wider move to frame the question of business and elite unity and conflict as an empirical question. Mizruchi (1989: 3) summarizes this core tenet of the neo-pluralist perspective:

> one fact has become abundantly clear: elites in advanced capitalist societies cannot be said in the abstract to be either unified or fragmented. There are times in which elites act in a unified manner and times in which they do not. What is needed, therefore, is a study of the conditions under which elites act in a unified manner; in other words, the factors that determine whether elites will cohere on a particular issue or series of issues.

Still, even if we depart from overarching concepts such as class and class interest, the question remains whether there are not certain business interests that tend to unite rather than divide business actors on given issues of contention. For it is entirely consistent with the pragmatic framing of this question as an empirical one to conclude that the possibility of such interest convergence needs to be taken into account, even if business unity cannot be assumed *a priori*. Several scholars of international environmental politics have indeed suggested that the business community's overall concern is to ensure that international environmental politics takes a 'business-friendly' course, and that specific environmental measures conform with a market-friendly model of regulation. This concern shows itself, for example, in business support for voluntary or market-based regulations (Eckersley, 1995) and a broader move over the last two decades towards a liberal environmental compromise that seeks to make environmental regulation compatible with the functioning of a liberal world economy (Bernstein, 2001). Thus, while international business groups may support the creation of multilateral environmental agreements to prevent unilateral state action, they will seek to limit the use of trade-related sanctions in MEAs so as not to disrupt the proper functioning of the World Trade Organization's (WTO's) trade system (Williams, 2001).

The cases studies in this book do indeed support the claim that business actors seek to make international regulation business-friendly. However, if this is not to be a truism, we need to specify more clearly how the business community seeks to make international environmental politics compatible with a supposed standard of business-friendliness or conform to the principles of a global liberal order. This, however, is a difficult, perhaps impossible, task, for it is not self-evident what business- or market-friendly regulation would consist of. Economists praise eco-taxes as the economically most efficient way of internalizing environmental externalities, but businesses usually oppose them. Leading European business groups, for example, have supported the Kyoto Protocol but at the same time resolutely opposed the introduction of a carbon/energy tax (see chapter 4). On other environmental instruments, such as emissions trading, one can find business groups on both sides of the argument, supporting and opposing such measures. Similar divisions exist in other areas, too. As Levy and Prakash (2003) argue, corporations on the whole support market-enabling multilateral agreements if they help to expand their global market reach, but other corporations that are concerned about suffering a competitive disadvantage will oppose them. In the same way, business groups often advocate replacing mandatory with voluntary environmental instruments, but heavily regulated industries may oppose the dismantling of regulations where they keep competitors from entering the market.

Thus, while business actors will most likely agree on the need to preserve a well-functioning capitalist world economy and to prevent unnecessarily burdensome environmental regulation, the *a priori* presumption of a uniform business interest is theoretically questionable and analytically unhelpful. The neo-pluralist perspective sees business power as a contingent concept and treats business interests and business unity versus conflict as empirical questions.

The place to engage with such questions, then, is in the historically bounded cases of international environmental politics.

II: Case Studies

3
Ozone Layer Depletion

The Montreal Protocol on ozone layer protection is frequently cited by politicians as the most successful international effort to protect the environment (Gore, 2006: 294). It is also considered by analysts as a clear example of the prominent role of business in environmental regime-building (Levy, 1997; Oye and Maxwell, 1995). Leading chemical firms, such as DuPont, Imperial Chemical Industries (ICI) and Hoechst, strongly influenced the negotiation positions of their respective home governments. Having initially opposed, and later supported, an internationally binding ozone regime, the chemical industry helped to make the Montreal Protocol a success by developing alternatives to ozone-depleting substances. As the case of ozone layer depletion suggests, the political economy of state–firm relations plays directly into the dynamics of global environmental politics. But what lessons does the Montreal Protocol experience hold for other environmental issues? Do corporations, because of their unrivalled economic power, possess a dominant, even 'hegemonic', position in this global policy area? Or is their power limited, because of divisions in the business sector and the mobilization of social power in global civil society? This chapter traces the evolution of business involvement in ozone politics from the origins of the ozone crisis in the mid-1970s to the mid-1990s. After setting out the environmental problem of ozone layer depletion and its origins in modern industrial processes, it reviews the creation of the global ozone agenda in the 1970s and early 1980s, the creation of the Vienna Convention and Montreal Protocol in the mid-1980s, and their evolution and implementation until 1995.

Ozone layer depletion and the CFC industries

Ozone is an atmospheric trace gas that is concentrated in a layer of the atmosphere roughly 10–50 kilometres above ground, the so-called stratosphere. It acts as a shield against ultraviolet radiation from the sun and thus plays a crucial role in protecting life on earth. Radiation from the sun sets off photochemical reactions in the stratosphere that continuously create and break down ozone. In the past, the natural cycle of ozone creation and destruction has produced fairly stable variations in the amount of ozone present in the atmosphere. However, recent increases in the atmospheric presence of certain trace gases, such as chlorine, bromine and nitric oxide, have accelerated ozone destruction. As we now know, the increase in these ozone depleters in the atmosphere is primarily due to the production and release of manmade chemicals (Makhijani and Gurney, 1995: chapters 1 and 2).

As soon as the thinning of the ozone layer was discovered in the early 1970s, scientists began to warn of the potential harm this could cause to human health and the environment. It was widely accepted that reduced levels of ozone in the stratosphere would allow greater amounts of biologically damaging ultraviolet radiation to reach the earth, and that this would in turn lead to a greater incidence of skin cancer, cataracts and immune disorders among humans. Just how many more illnesses and deaths this would cause was subject to intense scientific speculation. The US Environmental Protection Agency (EPA) claimed in 1987 that continued ozone depletion would result in 180 million new cases of skin cancers and 3.5 million cancer deaths by the end of the twentieth century in the United States alone (Litfin, 1994: 56–7). Laboratory experiments further suggested that plant life would be affected in a number of ways: increased ultraviolet radiation could lead to retarded growth of major crop species, thus threatening food security; it could reduce the production of phytoplankton in oceans, which forms the basis of many marine food chains; and it could contribute to a decrease in biomass production in terrestrial plant species, which might cause imbalances in the earth's larger ecosystems. As with human health effects, estimates of the likely impact on plant and animal life varied considerably, but few scientists doubted that continued ozone depletion posed a long-term environmental threat. The difficult, and politically explosive, question was whether the observed thinning of the ozone layer was the result of natural fluctuations in atmospheric ozone, or whether it was the result of manmade industrial processes.

The first scientific theory linking industrial chemicals with ozone depletion in the stratosphere was published in 1974 (Molina and Rowland, 1974). Based on laboratory experiments, two chemists from the University of California at Irvine, Molina and Rowland, hypothesized that chlorine atoms contained in modern chemicals such as chlorofluorocarbons

(CFCs) could initiate a catalytic chain reaction with ozone molecules, causing a depletion of the ozone layer. Based on existing projections of CFC emission trends, they estimated that stratospheric ozone could be depleted by between 7 and 13 per cent by the year 2000. The paper caused an intense scientific debate, focusing on whether CFCs were indeed the cause of the chemical reactions that led to ozone depletion, how long CFCs would be active in the atmosphere, and whether there even was a long-term trend of ozone layer thinning. At first, it appeared that Molina and Rowland had merely set the stage for a long and protracted scientific debate. But their suggestion that a ubiquitous group of chemicals might pose a global environmental threat was quickly picked up by the nascent environmental movement and regulatory authorities in the United States (Roan, 1989: chapter 2).

From an environmental perspective, the answer to the ozone depletion problem was straightforward: take precautionary action by reducing, and eventually eliminating, emissions of ozone-depleting substances (ODS). But the economic side looked more complicated. ODS such as chlorofluorocarbons were widely used in society, from refrigeration and air conditioning to cosmetics and electronics, and were a key ingredient in many industrial processes. In a sense, CFCs symbolized both the remarkable success of industrial chemistry and its ecological dilemmas. For over four decades since their invention in the late 1920s, CFCs were considered completely safe chemicals that were originally introduced to solve the health and safety problems associated with earlier refrigerant chemicals such as sulphur dioxide and methyl chloride (Hounshell and Smith, 1988: 155). The early success of the 'miracle' chemical as a refrigerant spawned many new applications after the Second World War, and production of CFC-11 and CFC-12, the two most popular CFCs, rose year-on-year to reach an all-time high of 800,000 tonnes in 1974 (Chemical Manufacturers Association, 1991), just as the ozone depletion controversy started to cast a shadow over the future of this lucrative business.

The producers of CFCs were affected most directly by the ozone controversy. They also led the initial response and dominated the business community's lobbying effort through most of the international negotiations on the ozone regime. Two factors accounted for their dominant role: most CFCs were produced by only a handful of large, multinational, chemical firms; and as producers of the very substances that caused ozone layer depletion, these multinationals held the key to replacing CFCs with safe substitutes, at least in the eyes of most user industries and policy-makers. An oligopolistic market structure, combined with a high degree of control over technological innovation in this sector, gave the CFC producers a privileged position within the business community as well as in the international political process.

CFCs had been manufactured on an industrial scale since the 1930s. Production started in the US and gradually spread to Europe and Japan, but remained concentrated in the leading industrialized economies well into the 1980s. All but a handful of chemical firms controlled global CFC production. The American CFC market, by far the biggest in the world (44 per cent in 1974) was dominated by DuPont and Allied Chemical, which accounted for 42 per cent and 26 per cent of US sales respectively. Union Carbide produced 17, Pennwalt 10 and Racon 2 per cent (Gladwin et al., 1982: 85). After Union Carbide's departure from CFC production, DuPont's position became even more dominant, rising to close to 50 per cent market share in the US and around 25 per cent worldwide in the 1980s. European companies accounted for 33 per cent of world CFC production in 1974, of which two thirds was in the hands of only three producers: ICI of Britain, Hoechst of Germany and Atochem of France. Other significant centres of CFC production existed in Italy, the Netherlands, Japan and Australia, as well as the socialist countries of the former Soviet bloc. But based on market share figures, it is fair to say that five chemical firms (DuPont, Allied Chemical, ICI, Hoechst and Atochem) in four countries (the US, Britain, Germany and France) dominated global CFC production (OECD, 1976: 28–9; US House of Representatives, 1987: 406–12). From the beginning of the ozone controversy in the 1970s until the Montreal Protocol negotiations in the 1980s, the structure of CFC manufacturing changed very little, with only a few developing countries emerging as hosts to new but small CFC-producing firms. CFC production was a mature business that did not attract many newcomers. The only noteworthy change in the industry during this time was the shift in market share from the US to Europe and Japan, partly in response to the stronger environmental concerns and regulatory pressures in North America. While the US share of the global market fell from 44 per cent in 1974 to 28 per cent in 1986, Europe increased its share from 33 to 42 per cent, and Japan from 8 to 12 per cent (Rowlands, 1995: 105).

The users of CFCs comprised a much more diverse and large set of companies, ranging from small retail businesses to large industrial manufacturers and the military. The sheer magnitude and diversity caused the user sectors serious collective action problems and proved to be a major disadvantage in the international political process. Due to organizational and financial constraints, most small companies restricted their political lobbying to the national level and acted mostly in a reactive manner. Even larger user firms relied on the CFC producers to represent their interests in the international negotiations. Users thus were dependent on the political and commercial strategies of the chemical industry, but because they represented a wider spectrum of the business community and accounted for the majority of jobs threatened by a switch away from CFCs, user firms possessed latent lobbying power particularly at national level.

Over time, the relationship between CFC producers and users became more strained. Major differences in political strategy came to emerge as the leading producer firms signalled their desire to work with regulators in the creation of international CFC restrictions. Many, though not all, user industries were more reluctant to give up CFCs, and some suspected the CFC producers of trying to create a market for higher-priced CFC substitutes, which would leave the user industries bearing the main cost of conversion. Others, however, began to search for alternative technologies in order to eliminate CFC use altogether, thus threatening to undermine the chemical industry's effort to work towards a controlled and drawn-out process of CFC substitution.

Corporate responses differed across the various production networks around which the CFC business was organized. The *aerosol industry*, which used CFCs as propellants in spray cans, is composed of mainly small to medium-sized companies. Not all aerosol manufacturers were fully dependent on CFCs, but most used some CFC propellants for their products. The relatively small size of most aerosol manufacturers made the cost of switching to non-CFC solutions a problematic issue. But reputational risks at the consumer end of the production chain became a major driver for change, and since a diversity of propellants had been in use, the aerosol industry did not have to rely entirely on the CFC manufacturers to identify substitution strategies.

The *refrigeration and air conditioning industry* is also characterized by a production network that consists of a large variety of firms, ranging from large manufacturers of refrigeration and air conditioning equipment (e.g. York International, The Trane Company) to smaller firms distributing and servicing this equipment. A 1981 estimate suggested that the US industry alone comprised some 1,000 individually owned companies with around 3000 branch outlets (US House of Representatives, 1981: 8–10; see Air Conditioning, Heating and Refrigeration News, 1987a, for 1987 figures). Because the entire sector depended on CFCs as inputs for use in insulation and as a refrigerant, it worked more closely with the CFC producers throughout the international negotiations and supported their strategy of a controlled CFC substitution strategy.

The *foam production industry* was equally fragmented, with a mix of small, medium-sized and large companies. CFCs were used in the production of both flexible urethane foams and rigid foam production. The US Urethene Foam Contractors Association alone counted 700–800 member companies, with an additional 1,500 non-member companies, most of which were employing only 5–15 people (US House of Representatives, 1981: 27). The CFC *solvents* business was a relatively small and more specialized market, providing high purity cleaning agents and specialty chemicals for the electronics, biomedical, communications, aircraft, aerospace and defence industries. Although the solvent providers, which relied heavily on CFC-113, were small in number, they serviced a wide range of influential industry sectors. Ultimately, however, solvents made up only a small fraction of the overall production costs for companies such as Northern Telecom and IBM, and many sought to eliminate CFCs altogether rather than rely on alternative chemicals developed by the CFC producers. Other industrial uses of ozone-depleting substances, which will not be discussed in further detail here, include halons production for fire extinguishing equipment and CFCs used as sterilants.

Business unity proved to be of vital importance to corporate lobbying in the ozone negotiations. From the start of the ozone controversy, the CFC producers led the political campaign against national and international regulations. Within only a few years, however, the aerosol industry in the US broke ranks with the chemical industry and adopted a no-CFC policy for all its products. Other user industries kept up their opposition to regulation well into the 1980s, and many did so even after the chemical industry changed its strategy. As the discussion below will show, the structure of the production networks that tied producers and users together had a significant impact on the degree of unity that the sector could maintain. The refrigeration and air conditioning industry, whose business was heavily dependent on CFCs, represented one end of the spectrum, working closely with the chemical industry to shape the CFC phase-out programme. On the other end of the spectrum were the solvents manufacturers and their user industries, such as electronics producers, who became one of the first to completely eliminate CFCs from their business.

Creating a global agenda for ozone layer protection

The first business reactions to the discovery of a potential link between CFCs and ozone layer depletion were largely hostile. Following a pattern that can be observed in nearly all newly emerging environmental issues, industry leaders initially doubted the validity of the scientific claims and warned against hasty regulatory action. The global business community was particularly hostile to the suggestion that precautionary action should be taken against environmental threats even under conditions of scientific uncertainty. It therefore came as no surprise that, when Molina and Rowland's 1974 research paper received widespread attention in US policy circles, business groups attacked it as an alarmist and scientifically questionable theory.

The ozone crisis came at a difficult time for the chemical industry in the United States. Producers of CFCs and their users in the aerosol industry had just been through a difficult public relations crisis in the early 1970s, when consumers grew increasingly concerned that CFC-propelled aerosol cans might explode if exposed to high temperatures and that their fumes might lead to intoxication (Aerosol Age, 1975c: 16–18; Roan, 1989: 35). The 'exploding CFC cans' crisis eventually subsided, but rising environmental awareness was putting the industry's traditional self-regulatory model under strain. The creation of the US Environmental Protection Agency in 1970 and similar environmental agencies in Europe signalled an end to an era when industry set and monitored its own safety standards (Paterson, 1991; on the history of chemical industry regulation, see Brickman et al., 1985).

A three-pronged business response

CFC producers led the business response from early on and advanced three key arguments, focusing on the science, economics and politics of the ozone controversy. In public statements and submissions to policy-makers, chemical industry representatives stressed that the scientific basis of the ozone depletion threat was weak and inconclusive; that restrictions on the use of CFCs would be very costly and in some areas technologically impossible; and that if regulations were to be imposed, they needed to be agreed at international level to cover all major CFC markets.

The first and most important line of attack was on the scientific merits of the ozone depletion hypothesis. This was the dominant theme in the statements of industry during the first ozone-related Congressional hearings in the US (US House of Representatives, 1975; see also DuPont, Inc., 1979: 6). While trying to discredit what it saw as an unfounded ozone scare, the chemical industry nevertheless called for further research to determine whether a link existed between CFCs and the thinning of the ozone layer. This was entirely consistent with the chemical industry's research-focused culture and business model that depended greatly on close cooperation with the scientific community. Soon after the ozone controversy erupted, the American chemical industry set up its own research programme, handing over some $5 million to the CMA's Fluorocarbon Program Panel to support scientific projects that probed the link between CFCs and ozone layer depletion (European Chemical News, 1975). The intention behind this industry research effort was to undermine Molina and Rowland's theory, but as Litfin (1994: 64) argues there is no evidence of industry censorship of the work funded by the CMA panel.

Ultimately, the industry hoped that further research would exonerate CFCs and that the regulatory threat would disappear. DuPont, for example, tried to reassure its commercial customers that it did not consider abandoning the controversial chemicals as it believed that further research would show them to be safe (Aerosol Age, 1975e). In any case, more research would buy time and delay regulatory action, allowing CFC producers to continue marketing the controversial chemicals and to search for substitutes in the meantime. Insisting on full scientific proof had another, crucial, purpose. It was aimed at the idea that environmental regulations were justified, and indeed needed, even before science could conclusively demonstrate the existence of a specific environmental threat. Such precautionary regulation had been a longstanding target of industry lobbying, as it threatened to shift the burden of proof from environmentalists to industry, leading to a situation where the latter had to defend its products against what it saw as spurious scientific claims. The ozone controversy proved to be a critical case in which industry sought to prevent the discourse of precaution from entering environmental politics, but ended up reinforcing it when it decided to support regulation despite the absence of conclusive proof.

In the end, industry's insistence on the need for full scientific proof had unintended consequences that were to seal the fate of the CFC business. It helped to elevate scientific evidence, and with it the members of the scientific community themselves, to a central role in the international political process. As Peter Haas (1992) has argued, the epistemic community that formed around a group of atmospheric scientists played a key role in promoting a convergence of interests among the major CFC-producing countries. Ironically, the chemical industry reinforced the epistemic community's privileged position, and to some extent placed the future of CFC production in their hands. The leading chemical firms were themselves actively involved in the scientific research endeavour, and through the CMA's Fluorocarbon Program Panel were able closely to follow and interpret new scientific findings. The major CFC producers were thus part of the wider epistemic community, and their interpretation of the evolving scientific knowledge base was to play an important role in developing a consensus on international regulations.

Thus, once a scientific consensus began to emerge by the mid-1980s that ozone depletion was a reality and CFC emissions were a major contributing factor, industry was in a bind to accept the need for CFC restrictions. As DuPont's Raymond L. McCarthy famously stated in the first US Congressional Hearings of 1974, if 'credible scientific data developed in this experimental program show that any chlorofluoro-carbons cannot be used without a threat to health, DuPont will stop production of these compounds' (US House of Representatives, 1975: 381). Other CFC manufacturers followed suit and by 1976, Ralph Engel of the CMA could declare that all companies involved in the CFC business had vowed to stop producing or using CFCs if credible evidence demonstrated that they created an environmental hazard (Aerosol Age, 1976b). The question was when the tipping point would arrive at which the mounting evidence in support of the CFC–ozone depletion link was considered scientific 'proof'.

Not all companies were happy with the strong stance on scientific evidence that DuPont and other CFC producers pushed in the early phase of the controversy. The CFC user industries, particularly, were keen to stress more the economic importance of CFCs to the running of their business and to society at large, irrespective of whether the compounds are harmful or not. This second aspect of the business argument against CFC regulation, which focused on technological sub-stitutability and conversion costs, was somewhat overshadowed by the scientific controversy surrounding the Molina–Rowland theory, but came to play a more prominent role as mandatory CFC restrictions appeared on the horizon in the late 1970s. For many CFC user industries, CFCs had become a seemingly irreplaceable ingredient of their products (e.g. refrigeration, foam production) or production processes (e.g. electronics), and the economics of replacing them bore greater weight than the scientific argument. In contrast, CFC production was only a relatively small part of the overall business of leading chemical firms (2.6 per cent in DuPont's case; Gladwin et al., 1982: 85), and although they were keen to preserve this business for as long as possible, its loss did not pose a fundamental challenge to the chemical industry's business model. A gulf between CFC producers and users therefore existed from early on in the ozone debate, but gained political significance only during the 1980s when international CFC restrictions came to affect the wide range of CFC users (see 'Technological innovation and regime evolution', below).

Today, it has become commonplace to argue that the costs and technological challenges of phasing out ozone-depleting chemicals were comparatively minor. Indeed, the challenges and costs of combating global climate change easily dwarf those of solving the ozone depletion problem. But this is not the way it was seen when the ozone controversy erupted, and the view with hindsight risks understating the uncertainties involved in the process of creating international regulations and finding substitutes for CFCs. While cost estimates varied considerably during the 1970s, all business leaders agreed that replacing CFCs would be very costly to the economy. DuPont estimated the total retail value of the US industry involved in the production, packaging and distribution of CFCs to be $8 billion in 1974, and the total related employment to be approximately 1 million workers (Large, 1976).[1] In similar vein, Allied Chemical Corp. argued that a whole range of industrial production processes, from pharmaceuticals to petroleum refining, depended on CFCs as input factors, and painted an almost apocalyptic scenario in which CFC restrictions would throw the United States into dependency on foreign suppliers, affecting the 'balance of payments, inflation, and national defense' (US House of Representatives, 1975: 446).

The third major argument employed by business groups concerned the international dimension of regulations. All industry spokesmen strongly rejected unilateral regulation by the United States if other CFC-producing countries would not also enact similar measures, or if no international agreement to combat ozone layer depletion could be reached. They urged the US government to abstain from environmental leadership and, should regulations become necessary, convince other governments of the need to act in a coordinated manner. The call for international harmonization became a constant refrain of industry lobbying against CFC restrictions, and was clearly motivated more by a desire to reduce the political momentum behind unilateral measures than by a commitment to environmental multilateralism. Industry persisted in arguing that there was no case for regulation in the first place, and by urging for an international solution, it sought to weaken the anti-CFC campaign in the US.

Emphasizing scientific doubts seemed to pay off in the early days of the ozone debate. As the Molina–Rowland theory came to be scrutinized by other scientists, questions emerged that led the two researchers to acknowledge that the photochemistry involved in ozone depletion was more complicated than they had originally assumed, and that estimates of ozone depletion had to be revised downward (Rowland and Molina, 1976). As a consequence, the National Academy of Science, which was carrying out a major research assessment exercise on behalf of the US government, delayed the publication of its report for several months and recommended that regulatory decisions be postponed in light of the scientific uncertainties (Litfin, 1994: 65). The ensuing scientific debate raised further questions and produced contradictory findings, without fully invalidating or confirming the original hypothesis.

Given the lack of a scientific consensus, industry felt vindicated that further research was needed before regulations are enacted. However, this strategy depended crucially on industry maintaining a united front in its lobbying efforts. Although at first all CFC user industries had rallied behind the public pronouncements of the leading chemical firms, the commercial and political strategies of producers and user industries began to diverge. Dwindling consumer confidence in their CFC-based products compelled certain users to reassess their strategy. For them, the technical and economic feasibility of replacing CFCs became a crucial factor in deciding whether or not to oppose demands for CFC restriction – irrespective of the scientific merits of these demands.

First cracks in the business front

The industry that was to suffer most from the commercial fallout of the ozone controversy was the aerosol industry. Manufacturers of aerosol cans were singled out by campaigners and regulators for several reasons. First, due to their non-flammability and absence of unwanted odour, CFCs were popular aerosol propellants, making the aerosol manufacturers by far the largest CFC user industry in the mid-1970s. Around half of the CFCs produced in the US ended up in aerosol products, of which three-quarters were destined for the personal hygiene market. The percentage of aerosol use was even higher in Europe. Major CFC-producing countries such as Germany and France consumed 67 and 70 per cent of CFCs in aerosol uses, respectively (International Environment Reporter, 1980b).

Second, because aerosol products result in high emission rates of CFC propellants as the aerosol is discharged, curbing their use would have an immediate positive effect on the environment. Third, the use of CFCs as propellants in aerosol products was considered as 'non-essential', especially when compared to the use of CFCs as coolants or as cleaning agents. And finally, replacing CFCs in aerosol production was considered less problematic than finding substitute solutions in refrigeration, air conditioning and cleansing applications (Boville, 1979). For all these reasons, the aerosol industry was picked out as the first target of CFC restrictions, a choice that was reinforced by other CFC user industries when they argued for exemptions from regulation because their CFC use was of an 'essential' nature, unlike in aerosol uses.

At first, leading aerosol manufacturers organized a lobbying campaign to defend their CFC use. They had to rely heavily on information provided by the chemical industry (Aerosol Age, 1975b), but failed to create a united front among all aerosol companies, many of which did not rely solely, or even primarily, on CFCs. Despite the fact that the aerosol industry set up a Council on Atmospheric Science to lobby against CFC regulations, analysts complained that the industry's response was patchy, ill-funded and insufficiently effective (Aerosol Age, 1975a).

The most serious blow to the aerosol industry's lobbying effort was dealt by S.C. Johnson, one of the major producers of household apparel products in the US with a global operations and distribution network. The company ran full-page advertisements in US newspapers in June 1975, announcing that it would immediately cease using CFCs in its aerosol products in order to avert damage to the ozone layer (Roan, 1989: 59–60). The announcement took the industry by surprise and had a devastating effect on its anti-regulatory strategy. As Elizabeth Cook (1996: 8) argues, 'S.C. Johnson's high-profile move thus undermined the economic and technical argument and strengthened policy-makers' resolve to take precautionary action in the face of scientific uncertainty.'

S.C. Johnson's move so shortly after the announcement of the CFC–ozone depletion theory was motivated by a desire to position the company as environmentally friendly, but also signalled a more general concern among aerosol firms that the decline in consumer confidence in CFC products would hurt long-term business prospects. Indeed, market data for 1974 and 1975 already began to show a significant fall in sales of CFC-containing personal hygiene products (Aerosol Age, 1975c; IFC Inc., 1986: 7). This trend accelerated in subsequent years and led to the near total collapse of the CFC aerosol market well before the US government introduced a ban on CFC propellants in 1978. The official ban thus merely confirmed a market trend set off by consumer reaction and adjustments in the aerosol industry (International Environment Reporter, 1979a). The rapid transformation of the aerosol market was helped by the fact that, against the protestations of many industry lobbyists, substitutes for CFC propellants were in fact readily available. Some companies only used CFCs in a small part of their product range – S.C. Johnson was reported to have used CFC propellants in only 5 per cent of its products (Chemical Week, 1975).

The companies that were to suffer most from this market reaction were the CFC producers. Within only a few years, they lost a large part of their CFC market and saw the anti-regulatory business front severely weakened. Not only did aerosol manufacturers accept the need for precautionary CFC restrictions, they also made a virtue out of a public relations dilemma by advertising their newly redesigned products as 'CFC-free', thus further entrenching the public perception that these chemicals were hazardous. The chemical industry tried to avert the collapse of this market by recommending the use of alternative chlorine-based chemicals, such as CFC-22 and CFC-142b (Aerosol Age, 1976d, 1977). But to the dismay of the industry, most aerosol manufacturers simply eliminated CFCs by switching to non-chlorine substitutes such as hydrocarbons, none of which were made by the CFC producers. While the problem of flammability made hydrocarbon-based propellants an inferior choice in the 1960s, modifications in aerosol products and the lower cost of hydrocarbons served to seal the fate of CFC propellants by the late 1970s (Aerosol Age, 1975d, 1976c).

As the experience with the US aerosol industry shows, lack of business unity and changing perceptions of the limits to CFC substitution created favourable conditions in which environmental campaigners and regulators were able to push for a reduction in CFC emissions in the US. Critical factors such as the economic or technological hurdles to CFC reduction were not a given but changed in response to corporate strategy and market structure. A comparison of developments in the US and in Europe further illustrates this point. Alternative propellants, which came to dominate the US aerosol market by the late 1970s, were also available to European manufacturers. But against the background of weaker environmental and consumer pressure, the European industry chose to reduce, rather than eliminate, CFC propellant use.

In contrast to their American competitors, European aerosol manufacturers were opposed to replacing CFC-based products, which were popular with consumers. While the North American aerosol market saw a decline in the annual growth rate during the 1970s, European demand for aerosol products was rising robustly during this period (Aerosol Age, 1976a, 1975f). With an expanding market also came growing demand for CFC propellants, which, as market research suggested, was driven by consumers' preference for higher-quality CFC propellants. S.C. Johnson and other US aerosol producers that followed its CFC substitution programme experienced the difficulties of such a strategy when they saw their market share in Britain fall after they eliminated CFCs from their product range (Aerosol Age, 1978). There were also supply-side reasons for why complete CFC substitution proved more difficult in Europe. The European aerosol industry was more fragmented and decentralized than its US counterpart, and the many small and medium-sized manufacturers found it more difficult to absorb the high redesign and refitting costs of a switch to the more flammable and explosive hydrocarbon-based aerosol production.

All in all, no major European aerosol manufacturer switched over completely to non-CFC propellants in the late 1970s, and the industry as a whole lobbied European governments to oppose a total ban on CFCs. In order to deflect criticism by environmentalists, it reached voluntary agreements with governments to reduce the consumption of CFCs in aerosol products, but was able to assert that unique market conditions and technical hurdles made a complete CFC phase-out impossible in Europe.

An international regulatory gap emerges

The collapse of the CFC aerosol market and the announcement in 1977 of a nation-wide ban on non-essential CFC uses (to be effective by the end of 1978) set the US on a course of international leadership on ozone layer protection. At the invitation of the US Administration, UNEP held its first international ozone conference in Washington, DC, in March 1977. A year later, at a follow-up conference in December 1978, the EPA's deputy administrator, Barbara Blum, urged other nations to follow the US example of reducing non-essential CFC uses (International Environment Reporter, 1979a). The conference concluded with the recommendation that any reductions in the use of CFCs should, if possible, be achieved without binding regulations, a position that industry observers welcomed (European Chemical News, 1978). But EPA officials continued to work for more stringent measures, and announced at a further international meeting in Oslo in April 1980 that the US would introduce regulatory measures to freeze US production of CFCs at 1979 levels (International Environment Reporter, 1980a).

In the industrialized world, only a handful of CFC-producing or -using countries were sympathetic to the US approach. Canada, Denmark, Norway and Sweden, of which only Canada produced small quantities of CFCs, took similar steps to the US ban on non-essential aerosol uses, mostly through close cooperation with industry (OECD, 1981: 255–6, 276–80; International Environment Reporter, 1979b). But the major CFC producers of Europe remained sceptical about the US push for international regulation. France, the UK and Italy, which accounted for 10, 9 and 5 per cent of global CFC production in 1974 respectively, participated in the search for better scientific understanding of ozone layer depletion but argued against precautionary measures (Gladwin et al., 1982: 77; OECD, 1981: 258, 262–3, 283–6). Germany, the world's second larger producer of CFCs (11 per cent global market share in 1974), showed more sympathy with the American position. In response to pressures from environmental campaign groups, the German government had reached an agreement with industry in 1977 to make voluntary reductions in CFC aerosol use and to introduce environmental labelling for CFC-free spray cans (International Environment Reporter, 1979d). The positions of individual European governments were coordinated within the EU as early as 1978, but it was only in March 1980 that the EU Council adopted a target of a 30 per cent reduction in CFC propellant use by the end of 1981 and a production cap (International Environment Reporter, 1979c).

While the European CFC industry was, on the whole, content with the policy adopted by the EU and member state governments, US industry came under renewed pressure once CFCs had been phased out in aerosol manufacturing. The EPA announced its intention to extend CFC regulations beyond the 1978 aerosol ban at an international conference in Oslo in April 1980, citing new scientific findings contained in the 1979 National Academy of Sciences report (Roan, 1989: 102). In the so-called Advance Notice of Proposed Rulemaking of October 1980, the EPA laid out an ambitious plan to freeze, and later reduce, CFC production, relying on two regulatory mechanisms: a production cap coupled with mandated reductions in specific CFC uses, and a more innovative model of economic incentives based on permits, which would allow the market to decide on how to allocate the remaining CFC production across different uses. The contrast between US and European approaches could not be sharper. Whereas the EU adopted a modest cutback of CFC aerosol use largely in line with market trends, the EPA signalled that it was aiming at a more comprehensive reduction of CFC production and use, to force change in the industry. If the US regulatory proposals were to enter into force, they would increase the international gap in CFC regulation and, in the eyes of US industry leaders, give European firms a competitive advantage.

The public consultation process on the EPA's Advance Notice produced an unequivocal response from the business community. The EPA received over 2,000 written submissions, of which only four supported additional restrictions on CFCs (European Chemical News, 1980a). In a sense, the EPA proposal became the rallying cry for a beleaguered US CFC industry that had seen its share of worldwide CFC production slip year after year. CFC production in the US had been growing at a rate of 7 per cent annually since the 1960s but experienced a sudden and sharp downward turn after 1974. Between 1974 and 1975, US production of CFCs dropped by around 20 per cent, reducing the US share of worldwide production of CFC-11 from 50 to 35 per cent and of CFC-12 from 60 to 50 per cent. In the four years after 1974, US producers suffered a 40 per cent drop in CFC sales, leading to considerable overcapacity in North America (Gladwin et al., 1982: 65, 68). In contrast, European CFC manufacturers experienced a much smaller fall in sales, but were able to expand their share of the global CFC market.

Against the background of lax international regulations, the EPA's intention to extend CFC restrictions beyond aerosols posed a serious threat to the US chemical industry. CFC producers were able to shift some of their production capacity to other, hitherto unregulated, CFCs (European Chemical News, 1980b), but further regulatory pressures would threaten the industry's international competitiveness and long-term survival. Led by the chemical industry, the remaining US CFC sectors decided to unite in a major lobbying effort against further ozone regulations in the US.

The most visible sign of the newly found unity in business lobbying was the creation of the Alliance for Responsible CFC Policy in 1980. There was little doubt that the small group of US CFC producers, led by DuPont, were the driving force behind this new lobbying group. DuPont played a key role in providing scientific and technical know-how to the corporate lobbying campaign, and its 1979 dossier (DuPont, 1979) became a key source of information for smaller companies' lobbying efforts. But the Alliance was keen to stress that it counted over 500 companies among its members at the height of its campaign and represented the broad spectrum of CFC user industries. One key conclusion that the chemical industry drew from the experience of the late 1970s was that only a united lobbying effort between producer and user industries could avert further CFC restrictions. Such a broadening of the anti-regulatory campaign would divert attention away from the chemical producers and highlight the wider economic repercussions of a CFC phase-out.

The creation of the Alliance also reflected frustration with the Chemical Manufacturers Association and its Fluorocarbon Program Panel, which had to represent a much wider spectrum of chemical industry interests that held back its lobbying efforts and limited it mostly to national regulation. In contrast, the Alliance pursued a single issue campaign and became the leading industry grouping at international level. The Alliance continued the longstanding demand of the chemical industry that regulations ought to be based on a sound scientific basis. It stressed the need to avoid discriminating against any single industry should further ozone restriction become necessary. And it emphasized that multilaterally agreed regulation would be preferable to unilateral action.

The outlook for the CFC industry seemed to be improving in the early 1980s, although this was not simply the result of the reinvigorated business front. Ronald Reagan's victory in the 1980 US presidential campaign boosted the hopes of business for a change in the Administration's approach to environmental regulation (Business Week, 1980). Within days of taking office, the Reagan Administration made deregulation one of the key priorities in economic policy. The newly created Task Force on Regulatory Relief was to examine all existing regulation, and all agencies issuing new regulations were required to justify their action on the basis of a strict cost-benefit analysis. The impact of Reagan's new economic philosophy was felt immediately at the EPA. The agency's budget was cut by 10 per cent, and its staff fell from 14,075 in 1981 to 10,396 in 1982 (Vogel, 1988: 247–9). And during Congressional nomination hearings, the incoming EPA administrator Anne M. Gorsuch (later Burford) questioned the seriousness of the threat from ozone layer depletion (US Senate, 1982). Instead of pushing ahead with the Advance Notice, the EPA urged the US State Department to work towards an international solution for ozone layer protection (Roan, 1989: 104).

Encouraged by the change in the political climate, the chemical industry put a halt to its ongoing research into CFC substitutes. DuPont had reportedly spent over $15 million, and Allied Chemical $10 million, on CFC replacement research during the second half of the 1970s. Their work led to several potential alternatives but none was considered to be a perfect substitute (Alliance for Responsible CFC Policy, 1986a: 5). The problems were of an economic as well as technological nature. Many of the substances identified were found to be of inferior quality, and some required costly toxicity testing before they could be commercialized. They were also more expensive than existing CFCs and would in many cases require costly product and production process changes. Without consumer and regulatory pressure, no CFC producer was willing to introduce such chemicals that most user industries were expected to turn down (interview with Joseph Steed, DuPont Co., 10 April 1996).

By the early 1980s, it seemed as if the CFC industry had weathered the worst of the storm that the ozone controversy had caused. Although, the chemical industry had found it difficult to prevent the rise of the ozone agenda, its insistence on full scientific proof seemed to pay off as ongoing research produced greater uncertainty regarding the CFC–ozone depletion theory. The experience of the 1970s also demonstrated the importance of business unity in fighting the threat of regulation. The switch of the

US aerosol industry away from CFCs wiped out nearly half of the US CFC market, and CFC producers were therefore keen to tie the remaining CFC user industries into a united lobbying front. Their cause was helped by a lack of international consensus and more sceptical reactions in Europe, but as the events of the 1980s would demonstrate, they had won only a temporary reprieve.

Corporate strategy and the creation of the ozone regime

The CFC producers emerged from the 1970s ozone controversy having suffered considerable, though not irreparable, commercial damage. Declining consumer confidence dented the industry's overall growth potential, more so in the United States than in Europe, but a recovery set in as soon as the global economy moved out of recession and the controversy over ozone depletion receded to the background. New CFC applications in the electronics industry and growing demand for refrigeration and air conditioning installations led to a renewed expansion of CFC production in the 1980s. But the experience of the 1970s had alerted the chemical industry to the potential threats of the ozone debate and the dangers of business disunity in the face of consumer and regulatory pressure.

International efforts resume, 1982–84

The anti-regulatory backlash in the US and growing uncertainties regarding the CFC–ozone depletion hypothesis made international CFC regulation appear increasingly unlikely in the early 1980s (European Chemical News, 1981). In Europe and North America, the problems of acid rain and river and marine pollution dominated the environmental agenda, while the chemical industry faced closer scrutiny over its health and safety record as catastrophic accidents such as in Bhopal in 1984 dominated newspaper headlines. The wider public, therefore, took little notice of renewed efforts at the international level to put ozone politics back on track. In 1982, UNEP began to convene a series of expert meetings to lay the ground for a framework convention to protect the ozone layer (International Environment Reporter, 1982). Under the dynamic leadership of Mostafa Tolba (Benedick, 1991: 40–1), UNEP kept the international ozone debate alive even though the major CFC-producing countries, including the US, showed no inclination to move towards international regulation (International Environment Reporter, 1983a). At this stage, CFC producers on both sides of the Atlantic continued to pay lip service to their commitment to multilateral solutions while expecting little political movement in that direction.

Once again, domestic change in the United States provided the impetus for international progress. The resignation of Anne Gorsuch as EPA administrator in 1983, after a scandal-ridden and ineffective period of trying to scale back regulations, paved the way for a change in the EPA's ozone strategy under the new leadership of William Ruckelshaus (Vig, 1990: 36–41). Conscious of the strength of the business lobby, EPA officials argued in favour of a renewed effort to create a global ozone regime, based on the US ban on CFC aerosol use, rather than push for domestic restrictions (International Environment Reporter, 1984a). This strategy was also mandated by America's Clean Air Act that urged the agency to develop international standards consistent with domestic regulations. From October 1983 onwards, the US joined the so-called Toronto Group (Canada, Finland, Norway, Sweden) in calling for a framework convention together with a separate protocol to include specific CFC restrictions (International Environment Reporter, 1983b).

The new US negotiation strategy posed a dilemma for American industry. Given its past support for international solutions, it should have welcomed the new position. But the new dynamism that the US injected into the international process could lead to unintended consequences. The potential gains of creating a global level playing field could be lost if an international ozone regime ended up imposing new restrictions on CFC use in refrigeration and air conditioning, foam production and electronics. It is for this reason that US industry representatives criticized the new US position at meetings with government officials in November 1983 and January 1984 (International Environment Reporter, 1983c). They were particularly concerned that even a limited international ban on CFCs based on inconclusive scientific evidence would give a boost to the precautionary principle and might encourage the EPA to consider regulating other CFC uses at home (International Environment Reporter, 1984a). Still, the Alliance for Responsible CFC Policy came out in support of an ozone convention that would encourage the monitoring and sharing of relevant scientific and economic data, but without prescribing specific CFC restrictions (International Environment Reporter, 1983c). Promoting better scientific understanding had been a longstanding industry position, and European obstinacy was widely expected to stall any moves towards stringent CFC controls. As one industry representative commented in 1983, '[f]ortunately for us all, there's very little substance to the [US] proposal' (International Environment Reporter, 1984a).

European industries were even more opposed to international regulation, and kept a low profile internationally during the early 1980s. Both CFC producers and users did not want to go beyond existing European regulations that limited the overall production capacity and reduced CFC aerosol use by 30 per cent. They saw this as a necessary compromise that provided sufficient room to expand other CFC uses within the overall capacity cap. European industrialists also feared that the US push for international rules was designed to help the American CFC industries shift the competitive balance. Although US and European producers were cooperating in the CMA's Fluorocarbon Program Panel, a growing rift between them began to emerge that reflected the different regulatory pressures they faced in their home markets. European fears were further aggravated by the fact that the American industry had created a formidable lobbying organization in the form of the Alliance, while the Europeans lacked an equivalent body representing their interests in Europe and internationally (interview with Franz Nader, Verband der Chemischen Industrie, 2 July 1996).

The Vienna Convention on the Protection of the Ozone Layer, 1985

It was clear from the outset of the international ozone talks that the major CFC-producing countries would drive the regime-creation process. The US and the leading European economies together accounted for over 70 per cent of world CFC production and nearly 60 per cent of consumption (Oberthür, 1997: 30–1). They could veto any international accord, and lack of cooperation between them would render any agreement ineffective. Moreover, the major CFC-producing companies, and to a lesser extent the CFC-using industries, held the key to finding and commercializing CFC substitutes. The corporate strategies of the CFC industries and the positions of the leading states were thus closely intertwined.

A broad consensus existed early on in the talks that a framework convention should help coordinate research efforts and lay the ground for potential future regulations. But while the Toronto Group, including the US, advocated the signing of a protocol with specific CFC restrictions at the same time, the majority of the European countries, together with the Soviet Union and Japan, preferred a phased process that would allow for more conclusive evidence to emerge before binding CFC restrictions were adopted. The Europeans advocated a production capacity cap as a precautionary measure (International Environment Reporter, 1984c, 1984b; Benedick, 1991: 42–3).

As the negotiating parties staked out their positions between 1983 and 1984, it became clear that both US and European proposals reflected the conflicting interests of their national CFC industries and their national regulatory approaches (e.g. US Senate, 1986: 176). US representatives pointed out that a production capacity cap would allow European CFC manufacturers to expand their production due to existing over-capacity, which US CFC producers had eliminated by the early 1980s (Benedick, 1991: 43). The Europeans, on the other hand, accused the US government of simply seeking to export its ban on non-essential uses, which would put no further burden on its industry. European companies also complained that the US definition of non-essential uses was arbitrary, if not hypocritical, for it included aerosol products which were valued more highly by European consumers, but not air conditioning which Europeans then considered as a luxury good.

The preparatory meetings held under the auspices of UNEP failed to overcome the stalemate between the two major positions. Both were tabled at the March 1985 conference in Vienna, Austria, and after renewed failure to agree on a protocol with binding CFC regulations, the conference adopted the Vienna Convention on the Protection of the Ozone Layer, which merely stated a general obligation on all Parties 'to take appropriate measures [...] to protect human health and the environment against adverse effects resulting or likely to result from human activities which modify or are likely to modify the ozone layer' (Article 2.1). Crucially for the future of the ozone regime, the US and its allies succeeded in including a resolution which authorized UNEP to convene a workshop on scientific and economic aspects of the CFC–ozone depletion link, and to work towards reaching an agreement on CFC control measures by 1987. This provision introduced an important dynamic into future negotiations and paved the way for future efforts to tighten the regulatory screw (Benedick, 1991: 45–6).

Only a small number of industry associations had followed the Vienna Convention negotiations. They broadly welcomed the outcome, but knowledge of the emerging ozone regime was generally limited in the wider business community.[2] At this point, only the larger CFC producers were still engaged in the international process, giving them a strong position to influence business lobbying on ozone matters as a whole. The CFC producers also welcomed the outcome, which was not simply an opportunistic gesture. As Benedick (1991: 46) argues, US lobbyists played a crucial role in preventing the conference from collapsing at the last minute, when anti-regulatory members of the Reagan Administration sought to withdraw the US delegation's authority to sign the agreement. It was only after lobbying by the US chemical industry in favour of the convention that the delegation was allowed to go ahead with the signing of the treaty.

Although the Vienna Convention disappointed most environmentalists, the adoption of a convention-plus-protocol approach was to prove successful in the end. Skilful diplomacy and the evolution of scientific knowledge worked together to create the conditions for the adoption of the Montreal Protocol in 1987 that would introduce the first binding CFC restrictions. Explanations of this success story have focused on the role played by diplomats (Benedick, 1991), epistemic community (Haas, 1992) and discursive shifts in the regulatory politics of ozone layer protection (Litfin, 1994). But there is another side to the success of the negotiations after Vienna. Once a basic agreement on the need to act had been reached, negotiators needed to create a regulatory system that would not only maximize environmental benefits but also minimize economic costs. Quite how far negotiators could go with their regulatory proposals depended crucially on perceptions of their economic impact and technological feasibility. Just as uncertainty on the scientific basis of the CFC–ozone depletion link impeded diplomatic progress, so did uncertainty regarding the technological and economic dimensions. And it was with regard to the latter that corporations were in a privileged position to shape perceptions and country positions. The corporate response to the Vienna Convention would thus play a critical role in determining the pace and direction of the international process.

The adoption of the Vienna Convention had two major effects on the business community: it forced a re-evaluation of its political strategy and helped to divide CFC producers from certain user industries. One possible business response would have been to maintain the traditional position of supporting further research and rejecting CFC restrictions unless sound scientific proof had been found. This is essentially the position most CFC user industries took, which remained detached from the international process. A different, and more complex scenario emerged for CFC producers. Undoubtedly, the chemical industry wanted to preserve the CFC business for as long as possible, just as the users did. But this had to be weighed against the rising political costs of opposing regulations at a time when the momentum was growing for some form of international CFC restrictions. Multinationals such as DuPont, Hoechst and ICI had been particularly exposed to environmentalists' campaigns, and were already under fire for their role in widely reported industrial accidents (European Chemical News, 1985; Paterson, 1991). With CFC production accounting for only a small fraction of overall production, the ozone controversy might become just too damaging for the companies' reputation. A different strategy, one aimed at a gradual and managed switch to CFC substitutes, might therefore be the more sensible option. For the more costly substitutes to be accepted by the market, however, regulations would need to be introduced to change price structures and create incentives for substitute technologies.

The CFC producers change strategy

Shortly after the signing of the Vienna Convention, new research by the British Antarctic Survey team suggested major losses of stratospheric ozone over the Antarctic between 1982 and 1985 (Farman et al., 1985). This discovery did not conform to the expected pattern of ozone loss and created a major anomaly for atmospheric scientists. While it called into question existing model predictions and heightened scientific uncertainty, it also seemed to underline the urgency of international action to curb CFC emissions. The phenomenon that became known as the 'ozone hole' could thus have had two very different effects on international ozone politics, either weakening or strengthening the resolve of policy-makers to regulate CFCs (Maxwell and Weiner, 1993: 29). Indeed, business leaders were heartened by the fact that over a decade of ozone-related research seemed to throw up more questions than it could answer (Chemical and Engineering News, 1987a, 1987b). But public and policy-makers' reactions to the 'ozone hole' tended to emphasize the threat it posed to the environment and humanity. On Capitol Hill, US legislators expressed their concern at what they considered to be a turn for the worse (US Senate, 1986).

Although the parties to the Vienna Convention decided to ignore the ozone hole in subsequent negotiations due to the scientific uncertainties that surrounded it, its discovery nevertheless affected the international process. As Litfin (1994: 97) argues, it 'created a sense of crisis that was conducive to the precautionary approach eventually sanctioned in the Montreal Protocol'. Developments on the economic front also contributed to the growing sense of crisis. During the early 1980s, the chemical industry was able to argue for further research while CFC usage continued unabated because ozone depletion did not seem to reach catastrophic levels and CFC emissions remained on a low-growth trajectory. The discovery of the ozone hole, however, questioned the former assumption, and renewed buoyancy in the CFC market began to challenge the latter. Policy-makers needed to reduce both forms of uncertainty – scientific and economic – before they could realistically attempt to create international CFC regulations.

UNEP convened two workshops in 1986 to deal with both forms of uncertainty. The first meeting was held in Rome in May and revolved around the issue of compiling reliable data on global production, consumption and emissions of CFCs. The then only inventory of the global CFC market was produced by CMA, which compiled data for CFC-11 and CFC-12 in industrialized countries (in aggregate form), which accounted for around 85 per cent of global ODS production (Chemical Manufacturers Association, 1991). But the survey excluded data on CFC-113, the fastest growing segment of the market, thus allowing the chemical industry to argue that it had no plans to expand CFC production in the near future (Litfin, 1994: 87). Industry representatives rejected US arguments based on a study by IFC Inc. that a global ban on CFCs in aerosols was feasible and would result in net economic savings as it did to the US economy (IFC Inc., 1986; Alliance for Responsible CFC Policy, 1986b: 21; IAL Consultants Ltd, 1986). The Rome meeting ended without agreement on CFC market data and regulatory cost estimates.

The second UNEP meeting, held in Leesburg, Virginia, in September 1986, proved to be more successful. Participants discussed different regulatory options, and a general agreement on the need for international regulations emerged (International Environment Reporter, 1986; Benedick, 1991: 48–50), with industry representatives beginning to ask governments for clearer signals to guide the marketplace (Litfin, 1994: 92). The Leesburg meeting was to mark a turning point in international ozone politics, as became clear when DuPont announced a major shift in its political strategy four days after the UNEP workshop, which was to galvanize corporate support for international controls on CFC emissions.

As the world's leading CFC producer, DuPont was in a powerful position vis-à-vis policy-makers as well as the CFC industries. It had been closely involved in the international process from an early stage, and its own scientific expertise allowed the company to monitor closely the evolving scientific debate. In light of the changing dynamics of the international process, the company underwent a major review of its political strategy in the summer of 1986, after Dr Joseph Steed, who was considered to be more open minded about the ozone issue, took over as Environmental Manager of DuPont's freon division. The strategic review resulted in the decision not only to accept some form of international regulation but to 'publicly announce its new position and become actively involved in the policy-making debate' (Glas, 1989: 3). In other words, DuPont would not just accept the need for international regulation as a *fait accompli* but use its new strategy to shape the very process of setting international rules.

DuPont managers insist that the decision was based entirely on an assessment of the evolving scientific knowledge (Glas, 1988). Science clearly played an important role but, as the discovery of the ozone hole had shown, the scientific debate was still riddled with many puzzles, and no consensus was in sight on all the main components of the CFC–ozone depletion theory. In fact, DuPont's decision to support a CFC control regime was based on a *political* judgement that the time had come for precautionary measures given changing perceptions of scientific uncertainty. This decision was also influenced by an assessment of the economic side of the CFC market. Well before the 1986 UNEP workshops, DuPont managers had come to realize that CFC production was starting to pick up again and would soon return to its past pattern of long-term growth. This led senior managers to conclude in 1985 that the strength of the CFC market would eventually undermine the industry's lobbying effort against international emission limits (interview with Joseph Glas, DuPont Co., 9 April 1996). DuPont managers concluded that the pressure for international CFC restrictions would become overwhelming, and that it would be better for the company to try to engage in this process so as to shape the resulting regulatory regime.

DuPont's strategic shift had a direct effect on the chemical industry, first in the US and later also in Europe. Other CFC producers were still arguing that it would be sufficient to limit the growth rate of CFC production. In the end, DuPont prevailed, and the Alliance for Responsible CFC Policy announced on 16 September 1986, shortly after the Leesburg workshop, that it supported a global limit on the future CFC production capacity. Significantly, the Alliance's board of directors approved this statement on 4 September, four days before the beginning of the workshop (Reinhardt, 1989: 15; International Environment Reporter, 1986), suggesting that the industry was willing to go along with a capacity cap even before the crucial meeting in Leesburg.

With this policy shift, the chemical industry implicitly accepted a precautionary approach to the ozone problem, which proved to be of great importance in the run-up to the Montreal Protocol negotiations. It strengthened the EPA's and the State Department's pro-regulatory position within the US Administration and took the wind out of the sails of those government officials and industry lobbyists that continued to resist any form of international regulation. It also laid the foundation for closer cooperation between US industry and governmental representatives in the negotiation process, before and after 1987.

By taking a lead, DuPont and the chemical industry opened up the latent rift between CFC producers and users. Most of the user industries, just like the members of the US negotiation team, were taken by surprise when DuPont announced its new position. They had been consistent in their objection to any CFC restrictions and expected the CFC producers to lead the fight against them. Users felt they had nothing to gain from a shift to alternative substances and were on the whole unprepared for a phase-out of CFCs. Indeed, for years the CFC producers had insisted that suitable alternatives would take at least five to ten years to come to market and might not be found for certain uses. Now that the producers seemed to change tack, user industries feared being burdened with the additional costs of higher-priced substitutes and process changes. Some accused DuPont of showing a 'thin skin' and bowing to the pressures to environmentalists (interview with Paul Horwitz, EPA, 8 April 1996). For now, however, the CFC producers managed to move to a policy consensus within the Alliance to reflect DuPont's strategic shift.

The Montreal Protocol negotiations, 1986–87

The negotiations on the Montreal Protocol opened at a meeting in Geneva in December 1986, with delegates from 25 nations and an unprecedented number of industry representatives present. While the major CFC-producing countries agreed on the need to freeze CFC production and consider future reductions, sharp differences persisted between the US and Europe. The US delegation proposed to move towards the phased elimination of 95 per cent of all CFCs, while the EU was unwilling to go beyond a production cap and further study of the ozone problem, in line with the mandate agreed by European environment ministers (Jachtenfuchs, 1990: 265–6). Observers and US delegation members had no doubt that the European position 'followed the industry line' (Benedick, 1991: 68), particularly in light of British, French and Italian opposition to reduction targets. European governments did indeed consult and coordinate more closely with industry groups, and the French and other delegations included industry representatives (Litfin, 1994: 107). But divisions between European governments and industry groups soon began to emerge that would weaken the anti-regulatory forces in Europe.

A change in European attitudes became visible at the second meeting in Vienna in February 1987. The German delegation for the first time rejected the delaying tactics of the other CFC-producing countries and indicated Germany's willingness to take unilateral action (Litfin, 1994: 110). The new German position reflected two important shifts in domestic politics and the business sector. During the 1980s, the environmental movement was gaining in strength, particularly after the Chernobyl nuclear accident in 1986, and German CFC producers had begun to signal a more conciliatory approach, under pressure from environmental campaigns. Hoechst, Germany's major CFC producer, was the first European company in March 1987 to offer the recycling of used CFC refrigerants as part of a broader strategy to reduce CFC emissions into the atmosphere (Europa-Chemie, 1987a).

The shifts in Germany had wider repercussions in Europe. Under pressure from the German government, European environment ministers agreed in March 1987 that a 20 per cent reduction target for CFC production should be achieved within four years of the signing of the Montreal Protocol (European Chemical News, 1987b). French industry representatives strongly criticized the new position, and the European Chemical Industry Council was still hopeful of diluting the effect of Germany's new pro-regulatory stance (European Chemical News, 1987a). But while Britain's ICI acknowledged that it could live with the 20 per cent reduction target without supporting the new policy (European Chemical News, 1987b), Germany's Hoechst signalled full support for the EU position (Europa-Chemie, 1987a). The German chemical industry, having coordinated its efforts closely with the aerosol industry, went one step further in May by offering to cut overall CFC use by 20 per cent, but by 1988, i.e. several years before the EU's target would come into effect (Frankfurter Allgemeine Zeitung, 1987a). In fact, in August 1987, one month before the final Montreal Protocol conference, the German aerosol industry formally committed itself to the phase-out of all CFC propellants by the end of 1989. Accounting for about half of German CFC consumption, the aerosol industry's move enabled the German delegation to go even beyond the 20 per cent target proposed by the EU (Frankfurter Allgemeine Zeitung, 1987b; Europa-Chemie, 1987c).

In following DuPont's lead, the German chemical industry faced a similar strategic dilemma as its US competitor. By endorsing international regulations, Hoechst sought to assuage domestic environmental critics and internationally harmonize CFC regulation that might otherwise be imposed domestically but not on its competitors, particularly in Europe. German industry leaders were aware that such a shift in position would have a considerable impact in Bonn and a signal effect across Europe, and would provide an impetus for international negotiations (Europa-Chemie, 1987b). But it could offset a political dynamic that would be difficult to control, leaving the chemical industry on the slippery slope of ever stricter regulations. The chemical industry therefore needed to work closely with the German government to create a level playing field, in Europe and internationally, which is precisely what the German environment minister promised to do (Frankfurter Allgemeine Zeitung, 1987a).

By April 1987, a compromise proposal was beginning to emerge in the ozone talks, despite continued resistance by the British and French governments (International Environment Reporter, 1987; Benedick, 1991: 71–2). By June, industry observers were certain that a protocol with CFC restrictions was within reach (Air Conditioning, Heating & Refrigeration News, 1987b). A last-minute hurdle emerged in the US when the US Office of Management and Budget sought to hold back US approval of the compromise deal. But the last minute intervention by anti-regulatory forces in the US Administration could no longer count on industry support. Instead, the chemical industry lent its support to the US delegation, and the Montreal Protocol. 'In the final analysis', Benedick (1991: 64) remarked, 'industry preferred to face a stronger treaty, which would at least bind its foreign competitors, than unilateral U.S. controls with no treaty' (see also Litfin, 1994: 104–6; Doniger, 1988).

When the Montreal conference opened on 8 September 1987, it was clear that the outcome would be a compromise between the two main proposals of the US and Europe, and that this compromise would have more to do with balancing commercial interests than environmental considerations. As Kevin Fay of the Alliance for Responsible CFC Policy said at the signing of the protocol, '...this has now become less of an environmental issue and very much an economic one' (quoted in US Senate, 1994: 53). The outstanding issues still to be resolved concerned questions of the scope of regulation (which chemicals were to be covered), the form of regulation (production and/or consumption controls), the timing of regulation (including the base year for calculating future reduction targets), the use of sanctions against non-signatories (trade restrictions), the treatment of developing countries and the design of the treaty revision process (on the final phase of the talks, see Benedick, 1991, chapter 7, and Litfin, 1994, chapter 4). In almost all the compromises reached in these areas, commercial considerations played an important role. The basket approach to the variety of CFCs covered by the protocol allowed Japan, for example, to continue to expand CFC-113 usage in its booming electronics industry as long as it reduced other CFC uses, and the application of a combined production and consumption cap simply integrated the preferred US and European approaches. In the end, the parties agreed to reduce ODS by 20 per cent in 1993/94 and by an additional 30 per cent in 1998/99, with 1986 as the base year for calculating reduction targets.

The outcome, widely considered a landmark agreement in international environmental politics today, disappointed many environmentalists. It had given industry a generous transition period in which to prepare for reduced CFC usage. With the flexibility built into the agreement, certain CFC sectors did not even have to take any action before the end of the 1990s, as long as CFC reductions were achieved in other sectors. However, in exchange for a market-friendly reduction schedule, industry had made important concessions. By supporting the Montreal Protocol, it had agreed to a shift in the regulatory discourse towards precaution, and this would set a precedent for future revisions of the treaty. DuPont led this shift in 1986 when it decided upon a new political strategy. Its dominant position among the chemical producers allowed it to determine strategy for the entire sector, and the user industries reluctantly followed this line, expecting the CFC producers to come up with substitute technologies. As soon as the protocol came into existence, the technological uncertainty of a CFC phase-out therefore moved centre stage in the international process, giving business a privileged position in the treaty revision process. It is important to remember that the protocol was adopted under conditions of considerable technological uncertainty (Parson, 2003: 174–5). But the extent to which DuPont and the other CFC producers could use this technological uncertainty to their advantage and control the conversion and revision process depended crucially on continued business unity.

Technological innovation and regime evolution[3]

The successful conclusion of the Montreal Protocol negotiations in September 1987 owed a great deal to the shift in corporate attitudes and the emerging consensus between governments, industry and NGOs on the need for precautionary action. The question now was whether CFC producers and users would cooperate in implementing the international agreement, and whether the CFC restrictions could be strengthened in subsequent negotiations. In this regard, the strategies and investment decisions by the CFC industries took on an eminently political dimension.

Corporate reactions to the Montreal Protocol: the CFC producers

The first official industry statements left no doubt that the chemical industry was willing to work with the new ozone treaty, despite privately voiced reservations (European Chemical News, 1987c). Industry's declarations of support were not a mere public relations exercise. Instead of trying to sabotage the implementation of the Montreal Protocol at national level, the chemical industry and some user industries took practical, and even innovative, steps to implement the ozone treaty. In doing so, business actors lent critical support to the regime, but also ensured that they were closely involved in developing the CFC reduction scheme. Their technological power came to shape the international regulatory discourse after the Montreal conference.

The first evidence of industry's cooperative stance came in the form of an unprecedented move to pool the testing of CFC alternatives. In late 1987 and early 1988, two industry programmes were created to assess the environmental acceptability (Alternative Fluorocarbons Environmental Acceptability Study) and toxicity (Program for Alternative Fluorocarbon Toxicology) of alternative chemical compounds. The industry co-operatives counted among their founding members 13 CFC producers from around the world. While these activities were restricted to the non-competitive area of chemicals testing, some bilateral programmes went as far as combining research and development efforts, such as those between Kali Chemie and ISC, and Atochem and Allied-Signal (European Chemical News, 1989; Chemical Engineering, 1988b).

Beneath the surface of worldwide industry cooperation, however, the major CFC producers competed to capture the emerging markets for CFC substitutes. In contrast to the smaller CFC producers, large chemical firms such as DuPont, ICI or Atochem were already in a leading position. Not only did they have the necessary financial and organizational resources to invest in new technologies; they could also build on the experience of the 1970s, when the first ozone controversy prompted them to look into substituting CFCs. DuPont had spent some $70 million on substitutes research in the second half of the 1970s, and decided in 1986, before the Montreal Protocol was signed, to revive this programme. After the signing of the treaty, DuPont committed over $30 million in 1988 and over $45 million in 1989 to finding CFC alternatives – the largest of all research efforts undertaken in the late 1980s (Manufacturing Chemist, 1988; European Chemical News, 1989).

Technological power became a key source of influence for the CFC producers. As policy-makers looked to the chemical industry for technological solutions and never seriously considered governmental funding for research into substitutes, it was inevitable that the chemical industry came to define perceptions of the technological feasibility of CFC substitution. Their investment decisions assumed an important political dimension and became the critical link between international regulation and changes in CFC production and consumption patterns.

One of the first technological decisions taken was to develop chemicals that could easily replace regulated substances, so-called 'drop-in' substitutes. Finding functionally identical, or at least similar, CFC replacements was no easy task, though. After all, the commercial success of the original CFCs rested on their unique combination of non-flammability and low toxicity, and other chemical substances with lower ozone depletion potential would almost inevitably lead to higher toxicity or safety restrictions. For this reason, the chemical industry initially concentrated its research on close relatives to the widely used CFC-11 and CFC-12 (International Environment Reporter, 1988).

The leading contenders to replace the most common CFCs were hydrogenated CFCs – most notably CFC-22, later renamed HCFC-22 – and fluorocarbons without the ozone-depleting substance chlorine, so-called hydrofluorocarbons (HFCs). Due to their lower ozone-depleting potential, hydrochlorofluorocarbons (HCFCs) were not treated as regulated substances in the 1987 Montreal Protocol. They became a major component of the chemical industry's early substitution strategy. HFCs, which were thought to pose no threat to stratospheric ozone, promised a more long-term solution for replacing CFCs, and were being promoted primarily in refrigeration and air conditioning uses (e.g. HFC-134a). Despite continuing problems with toxicity and concerns about the contribution that HFCs made to global warming, a race soon unfolded to capture the emerging market for HFCs. In 1988, CFC producers were not expected to be able to produce HFC-134a on a commercial scale before 1993 (Manufacturing Chemist, 1988). But only a year later, DuPont announced it was leading the race to develop a manufacturing process, and said it would begin commercial production by the end of 1990, several months before ICI was expected to bring its first HFC-134a facility on-stream (ENDS Report, 1989). In similar fashion, the chemical firms rushed into expanding production of HCFC-22, despite warnings by scientists as early as 1988 that HCFC-22 might soon be considered an unacceptable substitute (Manufacturing Chemist, 1988). Industry and government officials in North America and Europe were keen to see 'drop-in' substitutes enter the market quickly, and thus largely ignored warnings that HCFCs and HFCs themselves might become the subject of future regulations.

While most CFC producers kept an open mind about potential substitute compounds, competitive pressures gave rise to at least two principal product strategies. The first group of producers, consisting of DuPont, Elf Atochem and Montedison, simultaneously expanded production of HCFCs and developed varieties of HFCs. This strategy was endorsed by the Alliance for Responsible CFC Policy, which warned policy-makers not to pursue hasty reduction schedules for these two types of transitional substances (Alliance for Responsible CFC Policy, 1989). In contrast, ICI moved more decisively into HFC production. The company initially increased existing HCFC production without adding further production capacity to meet short-term increases in demand, but decided in 1988 against HCFC-22 use for the aerosol market over fears of toxicity (Parson, 2003: 176). It sought to convince its customers and the British government that an early switch to HFCs was the most desirable conversion strategy (Jordan, 1997: 16–17). As a consequence of these two substitution strategies, US and European positions on the future regulation of transitional substances began to diverge, with the latter beginning to argue for more forceful reduction targets for HCFCs. The result of this was a near reversal of negotiation roles: from the early 1990s onwards, it was the Europeans, with the exception of France, that were leading the campaign for an early phase-out of the first transitional substitutes.

Corporate reactions to the Montreal Protocol: the CFC users

Unlike the chemical industry, most of the CFC user industries were not actively involved in the ozone controversy until after the Montreal Protocol was signed. Even then, many companies took their time to react to the CFC restrictions, assuming that either the CFC producers would develop alternative substances or policy-makers would leave sufficient time for them to adjust. Only a minority of user firms took up the challenge and initiated ambitious efforts to eliminate CFC use. The strategic choices made by the CFC user industries were to have an important impact on the evolution of the CFC phase-out regime.

The divergence in user industry approaches is striking, for there is little in the institutional design of the Montreal Protocol that can account for this.[4] Instead, we need to look at business strategies, market structures and corporate networks to find explanations for the variation in corporate responses, which, in turn, influenced the evolution of the international ozone regime. We can distinguish between three major factors that shaped the user industry response to the Montreal Protocol.

First, at a fundamental level, the nature of CFC usage influenced corporate strategies. In some cases, particularly in the aerosol industry, low technical barriers to substitution allowed for a relatively rapid conversion process. Second, the heterogeneity of the CFC user industries, ranging from small-scale refrigeration and air-conditioning service units to large-scale electronics manufacturers, accounted for a certain degree of variation in corporate responses. The user industries were too diverse to coordinate their approach to CFC conversion. Third, market structures and corporate networks between producers and users also played an important role. For example, where the CFC producers enjoyed a close working relationship with CFC users, as in refrigeration, they maintained a strong influence over the choice of substitute technologies by their corporate customers, which, in turn, gave them greater leverage in the treaty revision process.

The *aerosol industry* was the first CFC user industry to react to the Montreal Protocol. No significant technical barriers stood in the way of replacing CFC propellants. This was evident from the US aerosol sector, which had phased out ozone-depleting propellants by the late 1970s. While most European aerosol firms had successfully resisted CFC restrictions in the 1970s, the Montreal Protocol forced them to reconsider their stance. Initially, the European aerosol industry hoped that the chemical industry would come up with 'drop-in' alternatives. European CFC producers initially suggested HCFC-22 as the main substitute (ENDS Report, 1987), which required only minimal process changes. However, soon after the signing of the Montreal Protocol, splits within the European aerosol industry emerged that mirrored the US experience. As soon as individual firms, such as S.C. Johnson and Talbec, opted for non-ODS products that they were advertising as 'ozone-friendly', others felt they could no longer support HCFC-based solutions in a climate of heightened consumer awareness. The European aerosol industry switched *en masse* to hydrocarbons as the preferred alternative propellant. Although requiring higher initial investment costs to convert existing production plants, hydrocarbons turned out to be cheaper than HCFCs and had no negative impact on the ozone layer.

As a result of this market shift, the pattern of CFC propellant replacement in Europe closely followed the example set by the US in the 1970s. The producers of HCFCs had lost all leverage over the aerosol industry and were forced to write off a significant part of the HCFC market. By the end of the 1980s, most European countries were well on course to meet the EU-wide target of 90 per cent CFC reduction in the aerosol sector by the end of 1990 (UNEP, 1989a: 13). Developments in the aerosol sector thus sent strong signals to policy-makers in Europe in the run-up to the first treaty revision in 1990, enabling them to consider tougher CFC restrictions than had previously been envisaged.

The *CFC solvent user industries* reacted to the 1987 ozone treaty in the same reluctant and defiant manner as many other user industries. The electronics industry, the main user of CFC solvents such as CFC-113 and methyl chloroform, opposed the Montreal Protocol and had been arguing for some time that CFCs were essential to modern processes of electronics manufacturing (Chemical and Engineering News, 1988; Parson, 2003: 183). Their position was strengthened when chemical firms reported that finding a substitute would prove far more difficult in the case of CFC-113 than CFC-11 and CFC-12 (Chemical Engineering, 1988a). CFC-113 replacement was further complicated by the fact that the world's largest buyer of electronics goods, the US military, stipulated the use of CFC solvents. Initially, therefore, the user industry expected other actors, especially the chemical industry, to take the lead in the search for alternatives. It also placed its hope on persuading governments to extend the CFC-113 phase-out schedule, to allow for an economically painless conversion programme in a key industrial sector.

Three factors, however, brought about a change to the electronics industry's strategy, making it one of the first sectors to completely eliminate ODS. First, CFC solvents made up only a small fraction of the value of electronics products; they were part of the manufacturing process, not the end product. Second, the electronics industry did not consider its links to the chemical industry as essential to its business model. The chemical industry thus had limited influence over the CFC conversion process in this sector. Third, the success of the electronics industry was built on the ability to respond to changing market conditions with technological innovation. This suggested that the industry was more inclined to consider radically different solutions than those provided by the chemical industry.

The first efforts to find an alternative cleansing technology were made only shortly after the Montreal Protocol was adopted. AT&T and Petroferm announced in 1988 that a naturally derived product could be used to deflux electronic circuit assemblies, while other firms introduced process changes that made the cleansing of circuit boards redundant (Environmental Protection Agency, 1997: 68). As a consequence of these innovations, virtually all major electronics companies committed themselves to eliminating CFC use by 1995, and many reached this goal much earlier. At a time when the chemical industry was still searching for a replacement of the 'magical' CFC-113, most electronics firms had already embarked on a process of eliminating CFC solutions altogether. The effect of these initiatives was dramatic (Pollack, 1991). By 1992, worldwide CFC-113 consumption had fallen to 126,500 tonnes, down from 276,700 tonnes in 1988 (Makhijani and Gurney, 1995: 172). Within a few years, one of the most intractable cases of CFC usage had nearly disappeared from the international agenda, against all expectations of policy-makers and industrialists.

The *refrigeration and air conditioning industry* was one of the most reluctant users to respond to the Montreal Protocol. While other sectors managed to reduce their CFC consumption in the second half of the 1980s, the use of ODS in refrigeration and air conditioning went up, both as a proportion of overall global ODS consumption and in absolute terms, from around 420,000 tonnes in 1985 to over 480,000 tonnes in 1990. It was only in the 1990s that the sector began slowly to reduce consumption of these substances (Makhijani and Gurney, 1995: 132–3).

According to industry representatives, technological barriers stood in the way of replacing existing refrigerants. In the absence of a readily available 'drop-in' substitute, CFCs were seen as 'essential' to the proper functioning of residential and commercial cooling systems. However, other factors related to market structure and corporate strategy also played an important role. The close relationship between the chemical industry and refrigeration and air conditioning manufacturers prevented a more radical redesign of cooling systems. Instead, the refrigeration and air conditioning industry dragged its feet over CFC replacement and relied on the chemical industry to come up with solutions: initially, ODS with low ozone depletion impact, and later HFCs, particularly HFC-134a.

The CFC producers first offered HCFC-22 as the optimal substitute for CFC-12 refrigerants, leading refrigeration manufacturers down a path that would later complicate the complete phase-out of ODS. Only by the mid-1990s, in response to regulatory restrictions on HCFCs, did the refrigeration industry introduce HFC solutions for both refrigerants and insulation (Somheil, 1996: 29). But both these substitute choices were challenged by environmental campaign groups, who argued that an entirely different option – hydrocarbons – could replace existing technologies. Greenpeace, in particular, led an international effort to convince refrigeration manufacturers and consumers of the benefits of hydrocarbons as refrigerants, a technology that had already been developed but was rejected by manufacturers. Eventually, the Greenpeace campaign proved successful in a number of European countries, but failed to have an impact on the North American market.

Greenpeace's effort to introduce a CFC-free refrigerator model began in late 1991 in Germany and led to an agreement in June 1992 with DKK Scharfenstein, a near-bankrupt East German manufacturer, to produce ten CFC-free refrigerators in a pilot project (Ayres and French, 1996). Greenpeace subsequently used its campaigning clout to help market the new CFC-free refrigeration technology, dubbed 'Greenfreeze', in Germany and abroad. Germany's main refrigerant manufacturers initially opposed the campaign, but following a shift in the market and the reputational damage suffered by their anti-Greenfreeze campaign, they quickly adopted the new technology. By 1996, hydrocarbon systems were used in 90 per cent of the German household refrigeration market. The decision to replace HFC-134a with hydrocarbons dealt a major blow to the German chemical producer Hoechst and its efforts to expand HFC production as part of its substitution strategy (International Environment Reporter, 1993e).

Outside Germany, only a small number of countries adopted the hydrocarbon technology, among them Switzerland and the Nordic countries. In North America, the new technology failed to make an impact with domestic users. This was to a large extent due to US health and safety regulations and concerns about the flammability of hydrocarbon refrigerators as well as their lower energy efficiency. Moreover, the US refrigeration industry had come up with what it advertised as its own 'CFC-free' solution, using HFC-134a (which contributes to global warming) as a refrigerant and HCFCs (with a lower ozone depletion potential) for foam insulation (Cook, 1996: 6). The US industry had no incentive to reverse its technological choice and blocked Greenfreeze technology from advancing further in the North American market.

Despite the Greenfreeze setback, the chemical industry's HFC-based strategy largely paid off. The success of hydrocarbon systems was limited to Europe, and even there only to small-scale household refrigeration units. Large-scale commercial refrigeration systems, by far the largest market segment, continued to rely on the chemical industry's preferred substitutes, HCFCs and HFCs. Although the CFC producers failed to achieve exclusive dominance in the refrigerant substitutes market, they nevertheless secured the larger part of it. Thus, despite regulatory pressure, NGO campaigns and technological alternatives, the chemical industry's corporate strategy and producer–user networks were able to determine the CFC substitution path in this important sector.

Business power and the treaty revision process

How did the CFC substitution strategies feed back into the international political process? As discussed above, the chemical industry had signalled to negotiators its willingness to work with the Montreal Protocol. The CFC producers were confident that they could find substitutes for the two most important ODS, CFC-11 and CFC-12, but warned that finding substitutes for CFC-113 would prove more difficult (Chemical Engineering, 1988a). Given that the protocol called for only a 50 per cent CFC reduction over a ten-year period, governments and industry had reason to believe that implementing the ozone treaty was not too difficult. But corporate support for international regulation was more elusive when it came to environmentalists' demands for a complete phase-out of CFC production. The major CFC producers made it clear that their support for the Montreal Protocol was contingent on the adoption of a measured regulatory approach that would respect technical realities and commercial interests. A complete phase-out of CFCs seemed economically and technically impossible, particularly in the case of certain 'essential use' CFCs. Such questions of technological uncertainty came to dominate international ozone politics after 1987. Of course, growing confidence in the scientific basis of the ozone theory provided a major impetus for stricter CFC controls. But the greater focus on the technical barriers to phasing out CFCs played into the hands of business actors.

Less than a year after the Montreal Protocol, DuPont once again led the way towards a new business strategy. In March 1988, the leading CFC producer broke ranks with other firms by announcing that it endorsed the target of a complete phase-out of CFC production. DuPont's latest move, based on a unilateral strategic decision by the world's leading CFC producer, was to have a crucial impact not only on Washington's negotiating position, but played an important role in shifting the regulatory discourse towards the complete elimination of ODS.

DuPont took the decision shortly after NASA's Ozone Trends Panel published new findings on 15 March that raised serious questions about whether the Montreal Protocol's restrictions on CFCs were sufficient to protect the ozone layer. DuPont managers involved in the company's decision later attributed it to growing scientific evidence in support of further CFC restrictions (Glas, 1988) – a view that supports epistemic community approaches that emphasize the role of growing scientific consensus in regime evolution (Haas, 1992). However, DuPont's policy change was a strategic, and politically significant, decision. The NASA report did not present conclusive evidence in favour of the CFC–ozone loss theory, nor did it call for any particular policy response. DuPont and the other CFC producers could have insisted – as most of the user industries continued to do – that in light of the remaining uncertainties and economic costs of conversion the Montreal Protocol represented a reasonable compromise in the interest of precaution. Having taken the lead in ozone politics in 1986, however, DuPont saw the NASA report as signalling a *trend* in the scientific discourse that pointed in only one direction, towards a strengthening of the ozone regime. The logical conclusion was for DuPont to move ahead of the game and throw its weight behind a total phase-out goal. This, the company hoped, would make it easier to cooperate with policy-makers in designing an 'orderly' phase-out of CFCs. As the sequence of events following the Montreal Protocol demonstrated, the strategy was at least partially successful.

Corporate decisions also played an important role in determining Europe's response to the Montreal Protocol. The CFC-producing countries which had been most reluctant to agree to the ozone regime – Britain and France – continued to act as a brake on European decision-making after 1987. Having agreed to CFC reductions in 1987, the EU took over a year to translate the international treaty into community law (Jachtenfuchs, 1990). But within a relatively short period of time, Europe moved from a policy of foot-dragging to political leadership in speeding up the CFC phase-out. To be sure, domestic factors, particularly a strengthening of environmental campaigns across the continent, were an important factor. But given the closeness of industry–government links in Europe's corporatist environment, changes in corporate strategy played an equally important role. This was most prominently the case in Germany, which helped to overcome British and French obstinacy within EU institutions.

From late 1987 onwards, Greenpeace and other campaign groups stepped up their ozone campaigns and targeted Hoechst, Germany's biggest CFC manufacturer, as well as selected user industries, particularly aerosol manufacturers. As it turned out, Germany's aerosol industry was an easy target. It quickly broke ranks with other user industries and gave up its initial opposition to a complete CFC phase-out, having already agreed to a CFC reduction schedule in the run-up to the Montreal Protocol agreement. Other user industries, although reacting more slowly, followed suit and set the signals for a relatively early phase-out of CFCs in Germany (Frankfurter Allgemeine Zeitung, 1988).

The CFC producers' response was more cautious, but was soon followed by a major shift in strategy. In the first few months after the signing of the Montreal Protocol, Hoechst demonstrated good will by setting up the first European recycling system for refrigeration liquids containing CFCs, while remaining critical of calls for more stringent CFC regulations (Der Spiegel, 1988). In December 1988, in response to a parliamentary commission's call for a 95 per cent CFC reduction, Hoechst came out in support for an eventual phase-out of CFCs. Compared to other European CFC producers, Hoechst was in a strong position to follow through this new strategy. By late 1988, the company's own reduction programme was already three years ahead of the Montreal Protocol's second phase, and it was able to commit itself to a complete elimination of CFC production by 1999 (Süddeutsche Zeitung, 1988). The company eventually brought forward the phase-out date to 1994, becoming the world's first major chemical company to stop CFC production (International Environment Reporter, 1994d).

Having followed DuPont's leadership in 1988, Hoechst took the lead amongst its competitors in April 1989 when it went public with its new phase-out target of 1995, which was also adopted by Solvay, Germany's only other CFC producer. Even DuPont had only declared its intention to complete the CFC phase-out by 2000 (Wall Street Journal, 1989). The two German firms urged their government to support the development of substitute chemicals and to work for a European-wide harmonization of CFC reductions in order to create a level playing field. While Hoechst's move to some extent reflected the growth of the anti-CFC movement in Germany, pressure from environmental groups alone cannot explain Hoechst's strategic shift. After all, the decision to phase out CFC production had been taken well before Greenpeace's campaign reached its climax in the summer of 1989, when activists climbed onto cranes at Hoechst's Frankfurt production facility (Frankfurter Allgemeine Zeitung, 1989). At the time, the company was going through a major strategic change which saw higher-profit specialty chemicals promoted at the cost of low-profit bulk chemicals such as CFCs. Given that only 0.5 per cent out of a total annual turnover of around DM 40 billion resulted from CFC production (Der Stern, 1989), Hoechst managers were therefore keen to protect the company's reputation from the negative publicity surrounding its CFC business. To the dismay of Hoechst, however, the company found it difficult to capitalize on its leading position in the phase-out of CFCs and continued to be the target of environmental campaigns for years to come.

Encouraged by these developments in Germany's CFC industry, the German government adopted a national CFC phase-out plan which aimed for a reduction of CFC usage of 95 per cent by 1995, but stopped short of ruling out CFC production altogether. Germany then took this new policy to the EU level and advocated a 1995 deadline for the whole community. This was unrealistic, but sent a strong signal to the recalcitrant member states. The European Commission itself had proposed a 1997 target, while France, Britain and Spain argued for 2000 to be kept as the phase-out target. The EU eventually decided to aim for a global phase-out from 1997 onwards, and to ban the five regulated substances of the Montreal Protocol by 2000 (International Environment Reporter, 1989). Germany had succeeded in nudging the EU position into a more proactive direction. As in the case of DuPont, a strategic shift by a leading European CFC producer had paved the way for this policy change.

Revising the ozone treaty: from Helsinki (1989) to London (1990)

The decision by the EU to support a complete phase-out by the end of the century, and to push for an earlier date if possible, had an important signal function for the upcoming First Meeting of the Parties in Helsinki in 1989. Only one day after the EU had decided on its new position, President Bush announced that the United States would also phase out CFCs by 2000, 'provided that safe substitutes are available' (International Environment Reporter, 1989). Although the Helsinki meeting produced only a non-binding declaration, it laid the ground for a major revision of the Montreal Protocol at the Second Meeting of the Parties in London in June 1990.

In the run-up to the Helsinki conference, UNEP's assessment panels produced the first set of reports on the state of knowledge in the areas of atmospheric science, environmental impact of ozone depletion, and technological and economic aspects of CFC conversion. The Synthesis Report strongly suggested that the long stratospheric lifetime of CFCs made a wait-and-see approach undesirable. A complete and timely phase-out of all major ODS was, as the report put it, 'of paramount importance in protecting the ozone layer' (UNEP, 1989c: 28). This statement played an important role in strengthening the resolve of governments to revise the Montreal Protocol. It is important to note, though, that by that time the major CFC producers had already committed themselves to an eventual phase-out of CFCs. This corporate commitment was echoed by the 1989 technological assessment panel, which stated that:

> Based on the current state of technology, it is possible to phase down use of the five controlled CFCs by over 95 per cent by the year 2000 ... Given the rate of technological development, it is likely that additional technical options will be identified to facilitate the complete elimination of the controlled CFCs before the year 2000. (UNEP, 1989b: ii)

The UNEP technology review thus confirmed what industry insiders had known for some time: that technological innovation could significantly reduce the time needed to complete the CFC substitution process. This did not mean, however, that an international agreement on a revised CFC phase-out schedule was now within easy reach. Far from it, negotiations on the path and timing of the CFC elimination programme proved to be complex. What had changed, however, was that scientists, environmentalists and leading industrialists had forged a consensus on the need to close down the CFC business. In deciding the CFC phase-out schedule, corporate decisions on technological change were of paramount importance. The contracting parties recognized this by inviting industry experts to join UNEP's Technology Assessment Panel, which became an authoritative voice in the global regulatory discourse. Panel members were expected to act as technical experts and not as industry representatives. Nevertheless, involvement in the UNEP panels boosted the legitimacy of the CFC industries as partners in the search for global solutions, and gave them a much stronger position to shape the emerging regulatory discourse.

From a corporate perspective, the critical issues on the agenda of the Second Meeting of the Parties in London in June 1990 were the inclusion of other ODS, such as HCFCs, in the list of regulated substances; and the tightening of the existing CFC reduction schedule. In principle, all chemical producers were keen to safeguard their investment in HCFCs for as long as possible. They argued that if HCFCs were banned in the near future, investment in their production would be at risk and user industries would be reluctant to cut back on their current CFC usage until a safe and acceptable long-term alternative had been found. Governments on both sides of the Atlantic were initially sympathetic to these arguments, and in early 1989 the EPA gave assurances to US industry that the US delegation would seek to protect the use of HCFC-22 at the forthcoming Helsinki negotiations (Air Conditioning, Heating & Refrigeration News, 1989).

To the disappointment of environmentalists, the main HCFC-producing countries managed to block proposals for an early phase-out of the transitional substances. The final text of the agreement called only for the use of HCFCs to be limited 'to those applications where other more environmentally suitable alternative substances or technologies are not available', and to be ended 'no later than 2040 and, if possible, no later than 2020' (quoted in Benedick, 1991: 263–4). On the question of revising the existing CFC reduction schedule, the EU had put forward the year 1997 as the final deadline for eliminating CFC use, while the US and Japan preferred the year 2000. The difference in negotiating positions reflected differences in corporate interests. While Germany's Hoechst had declared itself ready for an earlier phase-out date, thereby undermining the more cautious approach adopted by ICI and Elf Atochem, the US producer DuPont remained doubtful about a complete phase-out in 1997 (Benedick, 1991: 171–2). In the end, the lowest common denominator position prevailed, and 2000 was agreed as the phase-out date.

Among the other outcomes of the conference, the EU achieved a concession that allowed its CFC producers to rationalize production Europe-wide. Environmentalists scored a victory by having methyl chloroform included as a regulated substance, against chemical industry lobbying at the conference. Reducing methyl chloroform usage (primarily as a solvent) promised the single most important short-run contribution to decreasing the rate of stratospheric ozone depletion (Litfin, 1994: 151). Crucial to the success of this proposal was the progress made by electronics firms in replacing ODS in cleaning processes, which undermined the lobbying effort by methyl chloroform producers. Indeed, UNEP's technology panel had concluded that 90–95 per cent of ODS use as solvents could be eliminated by the year 2000 (Litfin, 1994: 151).

Moving towards phase-out: Copenhagen (1992), Vienna (1995)

As Litfin (1994: 156) points out, 'two primary factors ... drove the treaty revisions, the scientific observations of unprecedented ozone losses and the rapid progress in generating alternative technologies'. In the aftermath of the 1990 conference, new scientific studies painted an even bleaker picture of the environmental damage that was being done to the stratosphere. At the same time, the second report of UNEP's technology and economic assessment panels in 1991 documented the rapid progress in finding CFC substitutes. The report predicted that by 1992, CFC consumption would be reduced to 50 per cent of 1986 levels, a target that the 1990 revisions had set for the year 1995. Furthermore, the report suggested that virtually all consumption of CFCs could be eliminated between 1995 and 1997 (Litfin, 1994: 164).

Given this optimistic outlook on the implementation of the Montreal Protocol, the next revision of the treaty, at the Fourth Meeting of the Parties in Copenhagen in November 1992, was widely expected to produce a further tightening of the CFC restrictions. Indeed, the Copenhagen conference agreed on the phase-out of most CFCs by 1996, along with carbon tetrachloride and methyl chloroform, while the phase-out of halons was to be achieved by 1994. Among the more contentious issues, the phase-out date for HCFCs was brought forward to 2020, although essential use exemptions were included in the agreement, reflecting strong lobbying mainly from the US chemical industry and refrigeration and air conditioning sector. The US delegation argued successfully that HCFC use in air conditioners for large buildings be permitted until 2030, to reflect the long investment periods in this sector, a position that received the support of France. To the dismay of environmentalists, the world's leading HCFC producers, DuPont and Elf Atochem, had succeeded in securing a largely industry-friendly outcome on the question of these transitional substances.

It became clear after the Copenhagen conference that the major regulated CFCs had reached the end of their lifetime. Industry in the developed countries had made sufficient progress for the phase-out deadline to be brought forward even further. The EU environment ministers decided only weeks after Copenhagen to move it to 1995 (International Environment Reporter, 1993a). This was followed by EPA's announcement of a proposed regulation that would ban most CFC uses by the end of the same year (International Environment Reporter, 1993b). The CFC producers had long stopped opposing shorter phase-out deadlines and concentrated now on securing a sufficient lifetime for their substitute chemicals. The announcement in June 1993 of an earlier phase-out for HCFCs in Europe, therefore, caused some concern on the part of European producers. But the European Commission's proposal to achieve a HCFC phase-out in 2014, rather than 2030 as agreed in Copenhagen, applied only to domestic consumption in the EU and did not stop European producers of HCFC to continue exporting the chemicals to other countries that still relied on the transitional substitute (International Environment Reporter, 1993c).

The firms that were most threatened by this development were not the CFC producers, but the user industries. While the major CFC producers and policy-makers in Europe and North America cooperated in seeking to eliminate CFC production at the earliest opportunity, some of the user industries that had delayed conversion efforts now faced a situation of rapidly dwindling CFC supplies. Their new problem was that America's CFC producers were planning to stop producing CFCs much earlier than anticipated – in the case of DuPont, in 1994.

It was the refrigeration and air conditioning sector that was hit hardest by the speed of the CFC phase-out. In the United States, car manufacturers reacted to the ensuing crisis by lobbying the government to grant exemptions under the 'essential use' rules that allowed prolonged CFC usage. The American Automobile Manufacturers Association, a powerful grouping of America's carmakers, pointed out that 140 million cars were in use that needed future supplies of CFCs to service their existing air conditioning systems. US carmakers were only planning to start introducing new systems with HFC-134a as a coolant in 1994, the very year that DuPont planned to stop manufacturing CFCs. As a consequence, the automobile industry faced a serious squeeze on CFC stocks in the near future, which in turn would result in higher servicing costs for millions of car owners (International Environment Reporter, 1993d).

Given the political sensitivity of the issue, the US government gave in to car manufacturers' pressure and approached DuPont with a request to extend CFC production by one year. The move proved to be highly embarrassing for the government and the chemical producers, as both had so far cooperated in speeding up the CFC phase-out. In the end, lobbying by the car industry and fears of a voter backlash against higher servicing costs won the day (International Environment Reporter, 1994a).

By the time the Seventh Meeting of the Parties was convened in Vienna in December 1995, the focus in international negotiations had shifted from the CFC phase-out programme to debates on international aid to developing countries (Falkner, 1998) and the inclusion of previously unregulated ODS, particularly the thorny issue of methyl bromide. On the question of the HCFC phase-out, industry's position had received a boost in the latest Technical and Economic Assessment Panel report published before the Vienna conference. The panel concluded that although technically feasible, an earlier phase-out date of 2015 would cause unjustifiably high economic costs, as existing HCFC-using refrigeration and air conditioning equipment would have a lifetime far beyond 2015 (International Environment Reporter, 1995h). The revisions of the HCFC regulations agreed in Vienna reflected the prevailing expert opinion. Although the phase-out date was officially moved from 2030 to 2020, a 'service tail' of ten years was included that allows industrialized countries to supply existing equipment with HCFCs. The decision was strongly criticized by environmental groups for reflecting industry needs rather than environmental concerns (Greenpeace, 1996: 4). Yet again, a coalition of HCFC producer and user interests prevailed in the negotiations.

Conclusions

The creation of the ozone regime is a clear example of how business power shapes outcomes in international environmental politics and of how business conflict opens up political space for progressive environmental policies. Leading chemical firms vigorously opposed CFC regulations in the early phase of the ozone controversy. However, against the background of mounting scientific evidence and political will, a strategic shift by DuPont led to the breakdown of the anti-regulatory business front and promoted the adoption of precautionary international regulations. The chemical industry was unable to prevent the emergence of a global ozone regime, but by engaging with the international process in a more cooperative way, it sought to influence the design and timing of the CFC phase-out schedule.

While the CFC producers dominated the business lobby in the agenda-setting and negotiation phase, CFC users assumed greater importance in the implementation and treaty revision phase. After the adoption of the Montreal Protocol, the interests of producers and users began to diverge more clearly, leading to a more rapid elimination of CFCs than expected by the producers. As the case of the CFC aerosol, solvent and refrigerant user industries shows, corporate decisions on technological innovation had a significant impact on the path and speed of CFC conversion. This provides a corrective to state-centric perspectives on the technology–regulation nexus that tend to view international regimes as technology-forcing (e.g. Parson, 2003: 193). Once corporate agency moves centre stage, the impact of different pathways of technological change on regime evolution becomes apparent. It is in this sense that corporate decisions on investment and technological innovation can be said to play a political role. In the context of the Montreal Protocol, technological innovation was essential to the effective implementation of the global CFC reduction plan, but could not be taken for granted. Quite in what way and how quickly the CFC industries would develop substitution technologies depended to a large extent on the decisions taken by both CFC producers *and* users. Regulations were an important factor behind this technological change, but the Montreal Protocol's regulatory framework alone cannot explain the pattern and speed of CFC conversion. Moreover, the process of CFC substitution in turn played into the treaty revision process, and perceptions of technological uncertainty shaped country positions and negotiation dynamics in the international process.

The extent to which the corporate sector was able to shape regime evolution depended on its ability to unite behind a common strategy. Such corporate unity, however, was an elusive entity in ozone politics. Differences in commercial interests and political strategies emerged not only between CFC producers and users, but also within the user industries. Where CFC user firms opted for the complete elimination of CFCs and substitute chemicals, as in the electronics sector, the position of the chemical industry, both in economic and political terms, suffered a serious blow. In contrast, where the user industries chose to rely on substitute chemicals provided by the CFC producers, as in the case of refrigeration and air conditioning manufacturers, the position of the chemical industry was strengthened in international ozone politics. In this way, the dynamics of competition and conflict among the major CFC industries reduced the sector's overall structural power, and the political agency of leading policy-makers and business leaders helped to overcome structural impediments to a phase-out of ozone-depleting substances.

4
Global Climate Change

Climate change is one of the most intractable environmental problems the world faces today. International discussions on how to respond to the threat of global warming started in the late 1970s, and the creation of an international regime to protect the ozone layer gave rise to hopes that the international community would come up with a similarly successful solution for curbing greenhouse gas (GHG) emissions. If phasing out ozone-depleting substances seemed difficult at the time, tackling global warming has proved to be by far the more complex environmental problem of its kind.

As in ozone politics, business has played a critical role in the international politics of climate change. A vast range of industrial sectors are involved in producing and emitting GHG emissions, and many different technological changes will need to be made to address the problem. Would business actors come to play an equally supportive role in international regime-building as they did in the case of the Montreal Protocol? At first sight, the same pattern of business responses seemed to repeat itself in climate politics: the first business reactions to the scientific discovery of manmade climate change were overwhelmingly negative, focusing on the uncertainties involved in climate science. As pressure grew to address the issue internationally, corporate representatives highlighted the costs of taking action and the need to avoid damaging the international competitiveness of firms and sectors. Slowly but steadily, a more diverse field of business interests and strategies has emerged, but powerful business actors continue to resist international climate action, even today.

Closer analysis reveals important differences between the politics of ozone depletion and climate change. The experience of the international negotiations in the 1990s has revealed just how much more complex and difficult it has been to find an appropriate regulatory response to global warming. Despite a number of technological innovations, no technical fixes could be found to replace fossil fuels, and progress in limiting GHG emissions has been much slower than in the case of ozone-depleting substances. The central role that oil and coal play in energy production and industrial manufacturing has limited the scope for rapid action – and has enhanced the veto power of recalcitrant business interests. Indeed, it would seem as if the fossil fuel industry's key position in modern industrialism is the central blocking force in climate politics.

Nevertheless, the patterns of business conflict and competition have started to change the dynamics of international climate politics. The political field has become more fluid today, and a range of new political alliances between business actors, leading states and environmental campaign groups have sprung up that seek to produce progress in reducing GHG emissions. Within the Kyoto Protocol and beyond, an increasingly pluralistic field of political activity has emerged, involving an ever greater diversity of business interests and strategies. This chapter traces the evolution of climate politics, and the role that business has played in it, from the 1980s, when first efforts were made to organize an international response, to the negotiation of the UN Framework Convention on Climate Change (UNFCCC) and the Kyoto Protocol in the 1990s, and more recent efforts to build on these efforts and create a more comprehensive system of global climate governance.

The science and business of climate change

Global warming – a global problem

The earth's climate has always been changing, but recent increases in the concentration of heat-trapping gases in the atmosphere have led to a pronounced and unusual rise in the global average surface temperature. This so-called 'global warming' effect cannot be explained as part of a natural change in temperature and is now widely regarded by scientists to be the result of human activities, in the form of GHG emissions. The global warming trend, which has been recorded for the last century, has deviated from naturally occurring fluctuations in the earth's temperature and is widely expected to continue into the future if current trends of GHG emissions remain unchanged. Scientists have attributed the 'enhanced' global warming effect, i.e. the rise in average temperature caused by human activities, to the burning of fossil fuels, changes in land use, and agricultural and industrial processes that emit the three main greenhouse gases: carbon dioxide, methane, and nitrous oxide (Houghton, 2004:

chapter 2). Concentrations of these gases in the atmosphere have risen steadily since the eighteenth century, when the industrial revolution began to spread from Britain to Europe and beyond. According to the latest scientific estimates, carbon dioxide in the atmosphere has increased from a pre-industrial level of about 280 parts per million (ppm) to 379 ppm in 2005, and methane concentrations have risen from 715 parts per billion to 1,774 in 2005. As a consequence, global average temperature has risen about 0.74 degrees Celsius and is predicted to rise by around 4 degrees, and possibly even 6.4 degrees, over the twenty-first century (IPCC, 2007: 2, 4, 13).

The relatively rapid increase in temperature over the next 100 years is likely to have severe consequences for many ecosystems, plant and animal species as well as human societies. Despite a general trend towards warmer average temperature, climatic conditions will vary considerably across the globe, and climate change will therefore have highly uneven effects in different parts of the world. Some regions will experience some positive effects, most notably in Northern Asia and in Canada where increased temperatures will increase agricultural yields. But most other regions and countries will suffer predominantly negative consequences, with the worst affected areas to be found in developing countries, which also suffer from restricted capacity to adapt to climate change. The negative impacts include rising sea levels that threaten millions of people living in coastal areas or in small island states; reduced crop yields due to greater environmental stresses; an increase in infectious diseases due to the spread of pests and pathogens; and greater water shortages in already arid regions (Houghton, 2004: chapter 7).

The science of climate change is extremely complex, and predictions of long-term effects are based on model calculations that are inevitably characterized by a high degree of uncertainty. The 1980s and 1990s saw a massive effort by the world's leading scientific institutions to develop a better understanding of climate change and its effects. Since the late 1980s, a panel of scientists appointed and funded by major governments has evaluated the evolving scientific knowledge base and has produced regular assessments of the scientific aspects of climate change. The various reports of the so-called Intergovernmental Panel on Climate Change (IPCC) highlight the uncertainties involved but also point to the emerging consensus on key facets of climate change research. Thus, the various IPCC reports from 1990 to 1995, 2001 and 2007 have used increasingly strong language to describe the conclusion that manmade GHG emissions are causing climate change, with the latest report stating that 'most of the observed increase in globally averaged temperatures since the mid-20th century is *very likely* due to the observed increase in anthropogenic greenhouse gas concentrations' (IPCC, 2007: 10).

The business of climate change

One important source of the complexity of climate change is the diversity of business sectors implicated in the emission of greenhouse gases. Virtually every major economic and societal function contributes to global warming, from transportation to agriculture, and from land use changes to industrial processes. In 2000, 77 per cent of global GHG emissions were in the form of carbon dioxide, which is the result of fossil fuel consumption and changes in land use such as deforestation and agricultural management. Methane accounts for 14 per cent and nitrous oxide for 8 per cent of global emissions, and both are the result of agricultural practices and, to a lesser extent, certain industrial and resource extraction activities (Baumert et al., 2005: 4–5).

The centrality of fossil fuel-based energy to the functioning of industrial society alone ensures that virtually every major business sector is affected by climate politics. At the supply side of the fossil fuel energy chain, companies that extract and process oil, coal and gas play an important part as providers of fossil fuels. These are then consumed in electricity and heat production, transportation, industrial manufacturing and other forms of fuel consumption. The companies involved in the fossil fuel production and consumption network are often referred to as the 'fossil fuel industry', mainly because fossil fuel-based energy forms a major part of their business operations (Newell and Paterson, 1998; Levy, 2005). The term is a useful reminder of the central role that fossil fuels have played in the rise of modern industries, from the start of the industrial revolution in the eighteenth century to the expansion of global industrialism after the Second World War. However, when considering the climate-related political strategies of individual companies and the technological challenges they face in reducing their carbon dioxide emissions, we need to disaggregate the fossil fuel industry to identify the differential effects of climate change policies on individual sectors or firms, and the growing diversity of strategies that they have adopted in recent years.

Indeed, closer analysis of business strategies in climate change reveals a complex picture of differential effects and diverging opportunities. For companies that produce oil and coal, the prospect of international restrictions on GHG emissions poses a fundamental challenge to their business model. Diversification into other energy sources with a lower greenhouse effect (e.g. gas) or no effect at all (e.g. solar energy, wind energy) is a strategic option that some (e.g. Shell, BP) have started to develop.

But none of these options can offer a business opportunity on the same scale as that of oil and coal production, and currently estimated reserves of fossil fuels are likely to sustain traditional oil- and coal-based business models for another 50–100 years. Climate-related policy measures will therefore have a direct and immediate impact on these companies, which is why oil and coal companies have been propelled to play a prominent role in international climate politics from an early stage.

Most energy producers such as electricity firms are dependent on a secure supply of energy sources. Since oil and coal still dominate energy production in most industrialized countries, most energy firms have therefore sided with oil and coal companies in international climate politics. Their overriding commercial interest is related to supply security and energy pricing, and they have therefore tended to view restrictions on carbon emissions as a threat to their business interests. On the other hand, as the pressure from environmentalists and regulators has grown more recently, electricity firms have expressed an interest in ensuring regulatory certainty and predictability, particularly as their investments in power plant capacity require a long-term planning horizon of several decades. Such concerns over the uncertain future of carbon restrictions has led some electricity firms to support a managed transition to carbon restrictions combined with efforts to develop renewable energy sources (Mufson and Eilperin, 2006). Provided sufficient alternative sources of energy can be developed, the corporate interests of fossil fuel providers and electricity firms can therefore be expected to diverge as climate politics progresses.

Industrial firms that use energy as a major input have also traditionally sided with the oil and coal industry, but have more recently distanced themselves from them. As consumers of energy, which in the case of steel, aluminium and cement production can make up a substantial proportion of overall production costs, these sectors have been concerned about the price impact of carbon restrictions. At the same time, manufacturing firms can benefit from efforts of reducing their energy consumption as part of a climate- and cost-saving strategy, and they can support efforts to replace fossil fuels with alternative energy sources as long as these result in secure and affordable energy supply (Van der Woerd, 2005).

Companies in the transportation sector are equally exposed to the impacts of energy price rises due to climate regulations. Freight companies, airlines and shipping companies are all price sensitive with regard to energy inputs, and have in the past opposed climate measures that would result in higher fuel costs. However, achieving greater fuel efficiency can serve the dual purpose of meeting emission reductions and cost savings.

Furthermore, many transport companies such as airlines operate at the consumer end of the fossil fuel chain and will therefore be more exposed to shifting environmental preferences of consumers. Some have recently signalled interest in carbon offsetting schemes and measures to reduce their contribution to global warming as part of a broader corporate social responsibility agenda (see case studies in Cogan, 2006).

Affected are also companies and sectors that do not directly belong to the category of the fossil fuel industry. Because they play no major role in emitting greenhouse gases, provide technological alternatives to GHG-emitting processes or suffer the consequences of climate change, these business interests have generally been supportive of international action to combat global warming. In this category can be found renewable energy firms (water and solar power, biofuels, geothermal energy) and the nuclear industry, as well as the insurance industry. The former are set to benefit directly from greater restrictions on carbon emissions in the form of increased demand for non-fossil fuel energy sources. The latter is faced with increasing insurance claims due to the higher incidence of climate- and weather-related catastrophes and would benefit in the long run from greater efforts to stop global warming (see case studies in Cogan, 2006). For these companies, climate change is not a business threat but a business opportunity. Quite how climate change policies and regulations benefit specific sectors or companies depends on a number of factors, including the mix of policy instruments and the geographic coverage of such regulation, thus making it difficult to determine *a priori* the strategic options and corporate preferences of firms that stand to benefit from climate action.

Whether climate change presents itself as a threat or opportunity to corporations is by far the most fundamental dividing line in climate-related business strategies. This can be seen most clearly in the different positions that oil and coal companies have taken, on the one hand, and the insurance industry, on the other. Between these extreme positions of opposition and support for international climate action is a wide range of business interests and strategies that present a more nuanced and complex picture. Most companies and industries fall into this middle ground, where different shades of grey characterize the business threats and opportunities they face in the changing dynamic of climate politics. Whether companies perceive international climate action as harmful or helpful to their business interest depends not only on the nature of the regulatory intervention but also on the competitive position of those companies within their industrial sector and their ability to produce or apply technological innovations. Membership in corporate networks and the wider political environment in which they operate influence the way in which these firms formulate their political strategies. A growing diversity of business strategies has indeed emerged as climate efforts moved from agenda setting in the late 1980s and early 1990s to the creation of the UNFCCC in 1992 and the adoption of the Kyoto Protocol in 1997. Variation in business positions can be observed between different sectors, between different regions and countries, and even between individual firms within a given sector. By the late 1990s, the business community displayed a bewildering array of corporate reactions that ranged from calls for drastic emission reductions to fundamental scepticism of the scientific basis for international regulation. This diversity in business responses, which in some cases has taken on the form of business conflict, has come to characterize business involvement in climate politics in recent years. The following analysis of international climate politics, from the early debates in the 1980s to the adoption and entry into force of the Kyoto Protocol and beyond, seeks to shed light on how business power and business conflict have shaped the outcomes of this highly contentious international process.

The rise of the global climate agenda and the UN framework convention

The first scientific studies on the link between global climate change and carbon emissions into the atmosphere go back to the late nineteenth century, but reliable evidence that the world is experiencing a prolonged period of global warming became available only in the second half of the twentieth century. Global atmospheric research in the 1960s and 1970s hardened the belief of some scientists that manmade emissions were causing a rise in average surface temperature, and in the 1980s climate change slowly emerged as a political issue on the international agenda. The first World Climate Conference held in Geneva in 1979 led to the creation of the World Climate Research Program, and at a 1985 meeting of meteorologists in Villach, Austria, scientists concluded that some global warming was now inevitable. As before, they called for more research but also took a more activist stance and suggested that an international agreement to restrict emissions be reached (Weart, 2003).

The focus of international debates on climate change during the 1980s was very much on clarifying the scientific aspects and establishing an inter-nationally integrated research effort. Policy-makers called on scientists to establish greater certainty about the likely future rise in global average temperature, the links between GHG emissions and climate change, and the potential impacts on human welfare and the environment. These debates took place against the background of the successful creation of a global regime to protect the ozone layer, a potential precedent for international climate action. The Montreal Protocol (see chapter 3) had benefited from the growing knowledge base on ozone layer depletion but was completed well before some important scientific uncertainties had been resolved. It therefore provided an important example of precautionary international action that was flexible enough to respond to the evolving scientific debate. Applying this experience to climate change, environmentalists argued for precautionary restrictions on greenhouse emissions that would then be re-negotiated and tightened as scientists came up with more solid evidence.

As the international debate on climate change gathered momentum in the second half of the 1980s, the question was whether enough political support could be found for precautionary action in the face of continued scientific uncertainty. Environmental campaign groups such as Greenpeace were calling for immediate action (Leggett, 1990), but virtually all major industries that depended on fossil fuel-based energy opposed such measures and questioned the underlying scientific basis for international regulation. While all industrialized countries were supporting climate research, most were as yet unwilling to impose demanding emission reductions on a united business front. Only slowly were governments in North America, Europe and Japan moving towards a consensus on the need for internationally coordinated measures.

One of the first critical innovations in international climate politics that was to have a major impact on climate debates was the creation in 1988 of the IPCC. Established by UNEP and the World Meteorological Organization, the IPCC's mission was to review on a regular basis the state of scientific research on climate change and provide policy-makers with summary assessments that could guide the policy-process. The IPCC organized its work in three working groups: on science; impacts, adaptation and mitigation; and on economic and social dimensions. Being an intergovernmental body, the IPCC has developed a reputation for the world's most authoritative assessments of climate-related research. Its work follows scientific review procedures but its conclusions are agreed by delegates that include the political representatives of member states. The assessments thus represent a broad scientific consensus but the wording of the summary reports is politically negotiated. Even so, the IPCC has managed to gain legitimacy as a review mechanism with an important signal function for climate politics (Oberthür and Ott, 1999: 3–12).

As in the case of ozone layer depletion, uncertainty existed not only in the science area but also with regard to the economic and technological options for reducing GHG emissions. Reducing carbon dioxide emissions from fossil fuels required a dramatic change in energy production and use across a wide range of industrial sectors. For any climate regime to work, therefore, required a considerable degree of participation and cooperation among leading industrial firms. That business would play a major role in shaping the regulatory discourse on climate change was beyond doubt. The question in the 1980s was whether enough pressure could be created to initiate change in what seemed then to be a united business front against climate action.

Indeed, major industrial firms from around the world warned against a rush into climate action before the main scientific issues had been settled. Nowhere was business opposition to the climate change agenda as virulent as in the United States. As Levy argues, the 'initial response by U.S. industry was rapid and aggressive' (2005: 81). The major oil and coal companies led the first business campaigns and assembled a broad-based alliance of anti-regulatory business interests, including major manufacturing firms. Shortly after the IPCC came into existence, over 40 corporations and business associations created the Global Climate Coalition (GCC), the world's first dedicated climate change lobbying group. The GCC represented primarily those sectors that were heavily dependent on fossil fuels, such as the oil and coal companies as well as car manufacturers. It counted among its early members leading firms and associations including Amoco, the American Forest & Paper Association, American Petroleum Institute, Chevron, Chrysler, Cyprus AMAX Minerals, Exxon, Ford, General Motors, Shell Oil, Texaco, and the United States Chamber of Commerce. Initially, the GCC was primarily designed as an American lobby organization aimed at influencing domestic climate politics. The presence of some US subsidiaries of European firms ensured a more global outlook, however, and as the international efforts to create a climate change regime developed momentum, the GCC re-oriented itself to become the premier industry lobbying group at the international level (Pulver, 2002: 61; Levy, 2005: 81–2).

Similar to the first business response to ozone layer depletion, the GCC emphasized the scientific uncertainties behind the global warming theory and demanded full scientific proof before mandatory restrictions on GHG emissions be adopted. It also provided funding for atmospheric science, but unlike in the ozone case, the industry effort was more aggressively aimed at undermining the theory that manmade carbon emissions were contributing to a long-term global warming trend.

Fossil fuel companies provided a platform for climate change sceptics that challenged the emerging scientific consensus. To influence public discourses, they also set up a short-lived PR group called Information Council on the Environment, which ran public relations campaigns that portrayed climate change as just 'theory, not fact' (Gelbspan, 1997: 34). Coal companies such as Western Fuels Association deliberately sought to prevent other energy firms from accepting the growing consensus on climate change and funded more sceptical scientists to challenge the global warming thesis (ibid.: 35–6).

The GCC's second line of defence focused on the costs of taking action against global warming. Highlighting the centrality of fossil fuels to American industry, the lobbying group warned in 1996 that restrictions on GHG emissions by 20 per cent 'could reduce the U.S. gross domestic product by 4% and cost Americans up to 1.1 million jobs annually' (quoted in Levy, 2005: 82). Virtually all major industrial sectors would be affected by climate change regulations, but those depending on fossil fuels as major outputs or inputs would be particularly hard hit. Closely related to this was the argument that taking regulatory steps in the US would harm American industry's international competitiveness. The relative economic costs of a proposed global climate regime were particularly worrisome to the GCC as it might not include all countries and competitor industries. Similar arguments had been made in the ozone debate, but unlike in the run-up to the Montreal Protocol negotiations industry groups did not promote the international harmonization of climate regulation as a way out of the competitive dilemma. Leading US business voices simply did not want to open the door to regulation, be it national or international. Moreover, US industries were more dependent on fossil fuel-based energy than many of their European and Japanese competitors, who had achieved greater energy efficiency and were leading efforts to develop renewable energy sources in response to the oil crises of the 1970s. Internationally harmonized climate regulation might burden US firms with higher adjustment costs and would fail to address the competitiveness concerns of American firms. Instead of pushing for international rules, US business groups argued for voluntary approaches that would leave the initiative in the hands of companies (Moore and Miller, 1994; Gelbspan, 1997: 100–1).

Industry's anti-regulatory arguments fell on fertile ground in the US. As the international preparations for the 1992 UN Conference on Environment and Development (UNCED) in Rio gathered pace, the US Administration had to grapple with the question of how to respond to European and other calls for a climate treaty that included specific targets and timetables for GHG emission reductions. The draft treaty being discussed in the run-up to UNCED called for a stabilization of emissions at 1990 levels by the year 2000. Although internal studies suggested that the US might achieve this target without incurring severe costs, key officials within the Administration were opposed to the idea of setting specific targets and timetables (Hopgood, 1998: 160). Despite the US's international lead in developing the ozone regime (see chapter 3), both the administrations of Ronald Reagan (1981–89) and George Bush (1989–93) favoured a broadly deregulatory approach to environmental matters. Against the background of the global economic recession of the late 1980s, which had hit the US economy particularly hard, the Bush Administration was receptive to corporate arguments that climate regulations would harm the economy, reduce industrial competitiveness and threaten US jobs (Paarlberg, 1997).

Does US obstructionism in the run-up to the UNCED meeting in Rio suggest that business power reigned supreme in US politics? At first sight, this seems to be a legitimate interpretation, for as the end of the preparations of the Intergovernmental Negotiating Committee (INC) approached, it had become clear that the US Administration was as sceptical about binding GHG reductions as was the corporate sector. Against the wishes of environmental campaigners, the US delegation's mandate for the preparatory UN meetings was restricted to support for a weak framework convention without targets or deadlines for reducing fossil fuel use (Hatch, 1993: 13; Hopgood, 1998: 155–68). It was clear to many observers that the US delegation, as Hildyard put it, 'faithfully reflected the position of the oil industry' (1993: 29). But this assessment represents a frequently made but problematic conclusion regarding the power and influence of business. There is no question that the positions of the Bush Administration and the fossil fuel industry were closely aligned, but whether the latter influenced decision-making in Washington remains unclear. For the Administration's core team was already ideologically committed to an anti-regulatory agenda, and did not need to be convinced by industry lobbyists that a global regime to restrict carbon emissions would run counter to its policy preferences. Fossil fuel lobbyists found powerful allies in White House Chief of Staff John Sununu and John Dingell, the chairman of the Energy and Commerce Committee in the US House of Representatives (Hatch, 1993: 21). What the united business front against a climate treaty did provide was confirmation of the Administration's pre-existing conviction that it was defending the economic interest of the United States. Thus, the fact that the US delegation was given a closely circumscribed negotiation mandate for UNCED had as much to do with the ideological orientation of key actors at the heart of government as with a powerful fossil fuel lobby. Business was powerful and was heard in governmental circles, but did not determine US policy.

On the other side of the Atlantic, leading European governments and European Union representatives were slowly preparing the ground for an EU leadership position in climate change. In the second half of the 1980s, however, the first European policy initiatives were decidedly cautious. In its first Communication to the Council in 1988, the Commission merely stated: 'Reduction of greenhouse gases concentrations does not seem at this stage a realistic objective but could be a very long term goal' (Commission of the European Communities, 1988: 44). Two years later, however, the EU moved to a more proactive stance of advocating mandatory limitations on greenhouse gases for the industrialized world, by adopting a target of stabilizing GHG emissions by the year 2000 at 1990 levels. Little was known at this point about the potential economic repercussions such a move would have. The Commission felt it prudent to take a precautionary approach in light of the IPCC's 1990 report and was now considering how to support the non-binding target with concrete measures (Skjærseth, 1994: 26–7).

It was at this point that the EU moved ahead of what its business constituency was willing to support. Although the Commission concurred with industry groups that some emission reductions could be achieved by so-called 'no regrets' policies such as energy-saving measures, it also proposed to introduce a tax on carbon-based energy. This proposal clearly put the EU in a leadership position among industrialized countries and would have provided it with a strong position to push for strong measures and binding commitments at the Rio conference – a position that the Commission very deliberately sought to develop (Commission of the European Communities, 1991). But the Commission's carbon/energy tax was one step too far for the European business community. Despite having voiced more conciliatory views on the need for global climate policy than many of their American counterparts, European business leaders were adamant that the EU should not adopt policies that would harm European industry and undermine its international competitiveness. Leading industrial firms in Europe found it easy to mobilize a broad business front against the tax proposal, as most industrial firms would have been affected by such a tax, with only a small sector focused on energy efficiency improvements (e.g. insulation) and renewable energy sources potentially benefiting from it. European industry groups were thus able to present a fairly united lobbying front and put up one of the toughest fights against a European regulatory proposal – 'the most ferocious lobbying ever seen in Brussels', as *The Economist* commented (1992; see also Ikwue and Skea, 1994). In criticizing the tax proposal, UNICE, Europe's largest employer organization representing a wide range of industry groups, emphasized that a unilateral fiscal measure in the EU would run counter to the need for creating an internationally coordinated strategy (Skjærseth, 1994: 29). By implication, European industry conformed to the emerging consensus that international climate action was necessary, but was unwilling to bear the cost of European leadership in the form of unilateral regulation and advocated voluntary policy measures instead (Michaelowa, 1998).

Facing internal rifts between pro-environmental and pro-business Commission directorate-generals and unease among Southern member states, the EU settled with a compromise proposal for a carbon/energy tax that was conditional on the adoption of similar measures in other Organization for Economic Cooperation and Development (OECD) countries and contained several exemption clauses. The latter gave European industries with high energy consumption a partial exemption and allowed member states to suspend the tax temporarily (Skjærseth, 1994: 30–1). The compromise was agreed at the eve of the Rio Summit, and the EU thus entered the final round of talks with its leadership ambition significantly curtailed by an alliance of recalcitrant member states and powerful business lobbying. When in early 1992 EU Commissioner for Environment Ripa di Meana criticized the US for refusing to commit to specific emission levels or targets, it was still unclear whether the EU could live up to the political rhetoric it had brought to the INC negotiations (International Environment Reporter, 1992c). Still, given the more sceptical line taken by Japan (International Environment Reporter, 1992d) and outright US opposition to mandatory reductions, the European Union provided the main impetus for a strong convention on climate change.

The question of mandatory targets and timetables had been the central point of contention in the pre-UNCED climate talks. The idea that emission reductions were needed to stabilize GHG concentrations in the atmosphere was first put on the international agenda in 1988, at an international scientific conference on climate change in Toronto. In the so-called Toronto declaration, delegates warned of the far-reaching consequences of global warming and called for a long-term carbon emission reduction target of 50 percent, with an interim target of a 20 per cent reduction by the year 2005 (Newell, 2000a: 56–7). Proposals to include a reference to the Toronto target in a draft declaration at the Second World Climate Conference in 1990 were blocked, however, by the United States, the Soviet Union and Saudi Arabia, after heavy lobbying by the fossil fuel industry (ibid.: 108–9). This set a poignant precedent for the first meetings of the INC, the body set up in 1990 by the UN General Assembly to produce a draft treaty on climate change for adoption at UNCED.

The climate talks were also plagued by continued uncertainties regarding the scientific basis of global warming, which gave cause for caution on strong regulatory proposals. Shortly before the Rio conference, a science update produced by the IPCC for the forthcoming UNCED meeting scaled down estimates of global warming as a result of greenhouse gases and conceded that the observed warming effect could be the result of natural variability. Although confirming that industrial and human activities were increasing the amount of greenhouse gases in the atmosphere, the report also stated that the speed with which global warming is occurring might need to be revised (International Environment Reporter, 1992a). Moreover, research on ozone layer depletion suggested that the destruction of the ozone layer that had been caused by emission of CFCs had a cooling effect on the atmosphere (International Environment Reporter, 1992b).

For the anti-regulatory business lobby, failure to resolve such scientific debates underlined its insistence that the international community should abstain from mandatory targets for GHG emissions. The main policy recommendation that the business sector made in the run-up to UNCED was to focus on so-called 'no regrets' measures. These included attempts to increase energy efficiency, reduce energy use in manufacturing, and promote technological change and diffusion. In other words, such measures would be desirable in their own right and could thus be undertaken without damaging the interests of business and the global economy. Few business actors dissented from this position in a public way. The oil and coal industry played a dominant role in the pre-UNCED process and was able to rally a wide range of manufacturing firms behind its cause. Other business sectors were simply not involved in the process at this time or showed a greater degree of indifference, partly because the regulatory debate was focused on the major energy producers and user industries. Apart from the nascent energy efficiency and renewable energy sectors, no major global firm or industry spoke out in support of a global limit on GHG emissions (Grubb et al., 1999: 257).

The fossil fuel industry could count on support from the US, the other industrialized countries of the JUSSCANNZ grouping (Japan, Switzerland, Canada, Australia, Norway and New Zealand), as well as the countries of the Organization of Petroleum Exporting Countries (OPEC). The United States led the opposition to the idea of setting firm and legally binding targets, arguing that these would harm US industry and consumer interests and create global inequities. In line with industry's preference for 'no regrets' policies, the US delegation advocated 'improved energy efficiency and conservation, greater use of low emission energy and industrial technology, and improved agricultural practices', according to a US position paper leaked in February 1992 (International Environment Reporter, 1992b). In early 1992, US Environmental Protection Agency (EPA) officials were still trying to identify a level of emission reductions that could be achieved without significant costs to the US economy, in the hope of nudging the US position towards a compromise at Rio. This effort, however, ran up against the concerted opposition of anti-regulatory forces within the White House and industry groups such as the GCC, which argued that targets would threaten up to 600,000 jobs and a loss of $95 billion of gross national product per year (International Environment Reporter, 1992c).

In the end, all that the INC negotiations could produce was a compromise agreement based on the lowest common denominator. The final text that was adopted in May 1992 and opened for signature at Rio one month later committed parties to work towards the 'stabilisation of greenhouse gas concentrations in the atmosphere at a level that would prevent dangerous anthropogenic interference with the climate system' (Article 2, UNFCCC). In itself, this commitment was an important step forward in that it set a normative framework for subsequent efforts to deal with the climate change threat. Crucially, however, the framework convention did not specify the level at which atmospheric concentrations should be stabilized, nor did it include binding obligations to reduce greenhouse gases within a given timeframe.

Shortly after the draft text of the convention was released, representatives of the fossil fuel industry criticized it for setting a framework that acknowledged climate change as a future threat. According to Donald Pearlman, the executive director of the Climate Council, the text 'tries to lay a moral guilt trip' on industrialized countries (International Environment Reporter, 1992e: 254). But after the main elements of the treaty had been agreed, most industry representatives accepted the outcome and welcomed the fact that it did not set any binding targets for emission reductions or prescribe specific policy instruments (PR Newswire, 1992). Industry groups knew, however, that by creating a consensus on the need to address global warming and setting off a process of future negotiations on a protocol, the framework convention set the scene for an escalation of international climate action that could potentially run counter to industry interests.

What does the outcome of the UNFCCC negotiations tell us about business power and influence? That the Rio agreement on climate change took the form of a framework convention that excluded binding targets and timetables for GHG reductions can be seen as a major success for the business lobby. The vast majority of business actors involved in the talks had warned against mandated emission reductions. With the exception of the insurance and renewable energy industries, business was largely united in this. But despite this impressive show of unity, the fossil fuel industry was unable to prevent an international accord on climate change, as many in the industry would no doubt have preferred. As business observers had warned, the framework convention in their view set a dangerous precedent for a future tightening of international commitments. Business groups have always found it difficult to control the creation of new environmental agendas, and on this occasion had to accept that the momentum was growing for international action. The global environmental movement and progressive state leaders were able to define the agenda in ways that promoted a strengthening of the precautionary approach should the scientific evidence for manmade global warming further harden. For the moment, the reactive and largely obstructionist business approach worked, but there was no guarantee that this would continue to be the case in the future – and that the fossil fuel lobby could maintain a unified approach in a changing political field.

The Kyoto Protocol negotiations

The anti-regulatory stance of the fossil fuel industry played an important role in keeping binding emission reductions out of the UN framework convention. After the Earth Summit of 1992, the question was whether the business lobby could again block moves towards global GHG restrictions. The success of the fossil fuel lobby depended on a number of critical factors: the central importance of oil and coal energy to the industrial economies; the evolving discourse on the science of climate change and the technological hurdles to climate action; and the willingness of political leaders to press for more demanding regulations. But all these factors would be influenced by the degree to which the various industries dependent on fossil fuel-based energy could maintain a united business front. For if business divisions and conflict on climate change were to emerge, then the structural power of the fossil fuel industry would be reduced, and opportunities might emerge for a reassessment of the technological, economic and ultimately political feasibility of more stringent climate action. In this sense, business unity and conflict would become a key variable in determining business power and influence in the climate negotiations.

Emerging business conflict in climate change

Although the first splits in the business sector had already become visible at UNCED, they would only have a more significant impact in the run-up to the Kyoto conference in 1997. An important reason for this was the dominance of oil and coal companies among organized business interests in the first half of the 1990s. The companies that would be directly affected by potential climate regulations had been most actively engaged in the international process, and they remained in control of business lobbying particularly through the GCC (Pulver, 2002: 61). Other business groups more likely to benefit from climate action, such as the insurance industry and renewable energy firms, were either lacking in political and organizational clout or were too small and fragmented to counter the lobbying power of the major oil producers. For now, the most influential voices in the business sector remained opposed to mandatory restrictions on GHG emissions.

Over time, however, pressure for an agreement with specific targets and timeframes was building up. The UNFCCC included a mandate to negotiate a binding protocol with concrete measures, which provided environmentalists and political leaders with a platform to push for a more tangible climate regime. Their cause was helped by the IPCC's Second Assessment of 1995, which suggested that despite the remaining uncertainties the scientific basis for taking more drastic international measures was hardening (IPCC, 1995). Of course, the science of climate change was still evolving and many key issues had been left unresolved at Rio, but a process was now under way that would set the stage for a more focused negotiation of international commitments and timetables to act on climate change (Børsting and Fermann, 1997).

The first signs of a diversifying field of business interests and strategies could already be detected at UNCED. One of the outcomes of the 1992 summit was the creation of a new lobbying organization, the International Climate Change Partnership (ICCP), which took a more conciliatory approach to international climate action. Several large industrial manufacturers such as DuPont and Allied Signal had been actively supporting the Montreal Protocol on ozone layer depletion and were now turning their attention to the climate change challenge. They had gained considerable legitimacy from their involvement in ozone layer protection and were keen to distance themselves from the more aggressive anti-environmental stance of the GCC. Under the leadership of Kevin Fay, who had represented the Alliance for Responsible CFC Policy during the ozone negotiations, the ICCP put forward a more moderate industry position that recognized the threat of global warming and advocated a regulatory strategy that took into account the long lead times needed to find and adopt new technologies. ICCP's membership went well beyond the chemical industry and included a wide range of influential companies, including AT&T, Electrolux, Enron, General Electric and 3M (Giorgetti, 1999).

Divisions also emerged within the core group of oil and coal companies, albeit more slowly. The importance of fossil fuel-based energy to the global economy provided energy firms with a privileged position among all business groups, as has been widely noted in the political-economic literature on climate politics (Newell and Paterson, 1998). Whether and how this structural power would translate into political influence was shaped, however, by the degree to which it could maintain a united position. Losing the support of some of the major industrial energy users such as the ICCP's chemical and electronics firms was a significant, though not major, blow. More worrying for the sector was the growing rift that began to emerge between US and European energy companies. Whereas most American oil and coal firms remained opposed to any binding climate targets, Shell and British Petroleum (BP), Europe's leading oil companies, were beginning to take a more conciliatory stance from 1995 onwards. A Shell executive announced at the 1995 World Energy Congress that the world needed to start preparing for the orderly transition to renewable forms of energy while continuing to use conventional fossil fuels (Gelbspan, 1997: 86). And in October 1996, the American subsidiary of BP announced that the company was withdrawing from the Global Climate Coalition, in a move that signalled the deepest rift yet within the fossil fuel sector. The switch in strategy was confirmed in a high-profile speech by BP's chairman John Browne in May 1997, in which he acknowledged the growing scientific consensus on climate change, advocated taking precautionary action against it and announced a major investment initiative in solar energy (on the growing rift within the oil industry, see Boulton, 1997; Pulver, 2007; Rowlands, 2000; Skjærseth and Skodvin, 2003).

The most radical departure from the anti-regulatory business lobby occurred in a sector that was set to be one of the major losers to global warming. The global insurance industry had been aware of rising insurance claims due to extreme weather patterns since the early 1990s

(Hileman, 1997), and the world's largest reinsurers, Munich Re and Swiss Re, decided to take a more overtly political role in advocating progressive climate action to limit their exposure to rising insurance costs. Of all institutional investors, therefore, leading insurance companies came out most strongly in favour of a fundamental change in global corporate strategy. As early as 1992, both Munich Re and Swiss Re claimed that climate change posed the risk of bankruptcy for the global insurance industry (Schmidheiny and BCSD, 1992: 64–6). In 1995, 14 insurance companies from around the world signed a Statement of Environmental Commitment by the Insurance Industry, in which they committed themselves to a more systematic inclusion of environmental concerns, including climate change, into their risk and investment assessments (UNEP, 1995). By 2000, the number of the signatories to this UNEP-sponsored statement had risen to 84 (Paterson, 2001: 23).

These changes in corporate strategy had two positive impacts on international climate politics. First, they laid to rest the claim often made by lobbyists that the business community was opposed to mandatory emission restrictions, and that more ambitious emission reduction targets were economically and technologically impossible to achieve. As Michael Marvin, director of the Business Council for a Sustainable Energy Future, put it in 1996, the fossil fuel lobby 'wants you to believe that the science is divided, while business is not. In fact, the reverse is true' (quoted in Gelbspan, 1997: 85). After UNCED, more and more leading companies came to acknowledge the threat of climate change and the need for action, thus helping to shift the regulatory discourse into a more precautionary direction. Second, the growing diversity of climate strategies among major business actors opened up avenues for new political alliances between corporate leaders, NGOs and state officials in support of an international climate accord. The negotiations on the Kyoto Protocol would be the scene for a range of such initiatives that crossed the boundaries of previously opposed camps.

There have been important limits, however, on the effect that growing business conflict would have. On the whole, the pro-regulatory forces within the business sector are comparatively weak and lack the fossil fuel industry's richly funded organizational capacity. Even today, the renewable energy sector is dwarfed by the size of the fossil fuel sector and remains highly fragmented. In 2001, fossil fuels accounted for 64 per cent of world electricity generation, nuclear power and hydropower for 17 per cent each, and all other renewable energy sources for only 2 per cent (Sawin, 2004). The renewable energy sector is characterized by a large number of small and diverse firms that produce energy from wind power, water power, solar energy, biofuels and geothermal energy. These firms lack an overall organizational structure that would unite their business strategies and that could match the lobbying presence of the oil and coal sector.

A number of problems have also undermined the insurance industry's advocacy of international climate action. Although its role as a global investor in major industries endows the insurance sector with substantial structural power, it has not used this lever to bring about change in global energy production and consumption. This is partly due to the difficulties of shifting the insurance industry's large-scale share ownership, which is estimated to be between 20 and 25 per cent in major stock markets (Paterson, 2001: 19), out of the fossil fuel sector. After all, insurance firms derive 'much of their profit from investment activities in the very firms that contribute to global warming' (Salt, 1998: 164). A lack of sufficient alternative investment opportunities and the fiduciary duty to maintain shareholder value have therefore prevented a major shift in investment (Paterson, 2001: 31). Moreover, insurance firms have had limited success with the political initiatives they have injected into the international political process. Especially when compared with oil and coal firms, their lobbying effort has proved to be ineffective, held back by naivety and inexperience with the complex machinery of climate diplomacy (Salt, 1998).

Climate talks after UNCED: the 'Berlin Mandate' and beyond

Given the complexity of the climate change challenge and the centrality of fossil fuels to the functioning of the global economy, emerging business divisions in climate diplomacy were therefore unlikely to change international negotiation dynamics in quite the same way as in the case of ozone depletion. Still, the climate talks after Rio saw an unprecedented level of transnational lobbying and alliance building between states and nonstate actors, with business interests supporting different and conflicting positions in the regulatory debate. Environmental campaign groups worked closely with environmental regulators in the EU and the US, but were often frustrated by governmental departments representing economic interests. They therefore sought out allies among developing countries that would be most affected by climate change, particularly the small island states that saw rising sea levels as a threat to their very survival. Environmental campaigners and lawyers from the Foundation for International Environmental Law and Development, Greenpeace and the Worldwide Fund for Nature (or World Wildlife Fund) (WWF) worked closely with the Alliance of Small Island States (AOSIS), advising them on how best to represent their interests in the diplomatic process that is usually dominated by the industrialized countries (Newell, 2000a: 143; Arts, 1998). On the business side, similar alliances across national and regional boundaries emerged. The fossil fuel lobby, while influential in shaping the US position, also sought to build links with other states in the JUSSCANNZ grouping and especially with OPEC members, which had been playing an obstructionist role from the start of the climate talks (Newell, 2000a: 108–9).

The post-UNCED discussions on how to create international commitments on emission reductions revealed a large gap between the major negotiation groups, with the AOSIS delegations urging drastic measures to halt further global warming and many JUSSCANNZ and OPEC countries rejecting any binding targets and timetables for carbon reductions. While the AOSIS group was undoubtedly the most consistent *demandeur* for a far-reaching protocol, its limited influence and economic weight made it an unlikely leader in the diplomatic process. In contrast, the EU, which represented one of the world's major economic blocs and accounted for over 24 per cent of carbon emissions in 1990, was widely expected to move the negotiations forward and push for an agreement among OECD countries. The first opportunity for such progress came at the first Conference of the Parties to the UNFCCC (COP-1), which was held in Berlin, Germany, in early 1995.

The EU was committed under the framework convention to stabilizing carbon emissions at 1990 levels by the year 2000, and was arguing for further reductions beyond the 2000 deadline. Within the EU, Germany and Britain were leading the call for more ambitious GHG emission cuts. Their proposals aimed at reducing industrialized countries' emissions by 10 per cent below 1990 levels, to be achieved by 2005, and 15 per cent by the year 2020 (International Environment Reporter, 1995c, 1995b). At COP-1, Germany went one step further and committed itself to reducing emissions by 25 per cent by 2005 (International Environment Reporter, 1995e). Not all EU member states were willing to go along with these proposals, but a loose coalition of Northern European countries was able to set the agenda and promised flexibility in the setting of country-specific targets, which was eventually adopted as part of the EU's climate strategy in 1997 (International Environment Reporter, 1997a).

Several authors have noted the unique political economy of Germany's and Britain's climate policy (Eichhammer et al., 2001). Both countries were responding to rising environmental concern at home, which in Germany's case has led to the rise of the Green Party with parliamentary representation at federal and regional level. But domestic public opinion alone cannot explain their progressive stance. It is no coincidence that both countries have also benefited from a path of industrial and energy development that has made drastic emission reductions easier to achieve. In Germany, the inclusion of the former German Democratic Republic (GDR) in the national carbon emission figures allowed the country to aim for a higher reduction target than most other European countries without suffering immediate and high economic costs. The collapse of much of the GDR's industry after the fall of the Berlin Wall provided a major windfall for Germany's overall climate strategy. Britain, on the other hand, had since the 1980s reduced its dependence on carbon-intensive oil and coal-based energy sources by switching to North Sea natural gas, a structural change that was motivated by economic not environmental objectives. Both these structural changes enabled a push for higher EU and international emission reductions.

Both Germany and the UK also benefited from a more cooperative working relationship with key industry groups, including Europe's leading oil firms Shell and BP. In Germany, 17 industry associations joined in a 1995 pledge to a 20 per cent voluntary carbon emission reduction by 2005. The Federation of German Industries was keen to stress that no other manufacturing association in the world had made such a commitment, and expressed hopes that this voluntary offer would make mandatory targets in Germany unnecessary (International Environment Reporter, 1995b). Although the voluntary agreement between government and industry was criticized by opposition politicians for not going far enough, it nevertheless signalled a more conciliatory approach by leading German companies that stood in sharp contrast to the antagonistic stance taken especially by US industry groups.

By announcing an ambitious emission reduction target, the German government had hoped to set an equally ambitious agenda for the first meeting of the Conference of the Parties in Berlin in 1995. In the end, COP-1 was of rather limited success but managed to set the scene for the successful conclusion of the talks two years later in Kyoto in 1997. Opposition from the United States, Japan, Australia and the OPEC countries ensured that binding commitments to reduce emissions were not agreed in Berlin. But the conference concluded with a decision to set up a two-year negotiation process on a climate protocol, and this so-called 'Berlin Mandate' included a commitment by industrialized countries to take additional measures beyond 2000. As a concession to the US, the parties agreed on a programme for joint implementation that would allow industrialized countries to invest in emission reduction efforts in other countries but to be credited for this contribution to reducing GHG emissions (International Environment Reporter, 1995e).

The Berlin conference saw a huge turnout of industry representatives as conference observers. The growing split in the business sector between anti- and pro-regulatory positions was also in full show, as the GCC continued to oppose any move towards specific obligations while the insurance industry openly supported the AOSIS group's demand for a strong protocol. The middle ground was occupied by groups such as ICCP and the Business Council for a Sustainable Energy Future, which played a more constructive role but warned against hasty decisions on the timing of future negotiations and commitments. Overall, observers felt that the arrival of more moderate business interests at the negotiations had transformed industry lobbying (Dunn, 1995: 442). No longer could the fossil fuel industry appear to speak on behalf of the business community overall. Of course, most US oil and coal firms continued to take a hard-line stance and criticized the COP-1 outcome for failing to ensure participation of developing countries in any future reduction programme. GCC representatives chided US delegates for agreeing to the Berlin Mandate, which in their view went against previously made assurances that the US would not commit to any timetables. But the ICCP, which represented major companies such as DuPont and General Electric, supported the Berlin agreement and merely called on delegates to agree a process 'grounded in scientific, technical, and economic assessment' (International Environment Reporter, 1995f). Governments willing to push for binding targets no longer faced a united front of hostile business lobbyists, but could now draw more moderate voices into the process of assessing the technical and economic costs of climate action.

The domestic battle over US climate policy

In the new international political climate after the Berlin conference, leading US business opponents of climate action renewed their efforts to win Congressional support for their position as a way of tying the US delegation's hands in the climate talks. The more fluid the international process had become and the more the Clinton Administration warmed to the idea of binding emission targets, the stronger business groups such as the GCC and Climate Council focused their attention on the domestic level. Given that any protocol to the UNFCCC would have to be ratified by the US Senate, the climate sceptics among US business groups were in a strong position. Under Republican leadership, the Senate had grown increasingly hostile to international climate action. After the adoption of the Berlin Mandate, leading Senators such as Frank Murkowski were warning that the Senate would not ratify a protocol that in his view would harm US economic interests, while Republicans in the House of Representatives introduced legislative measures to cut back federal funding for climate-related research (International Environment Reporter, 1995g).

The oil and coal industry did indeed have reason to be concerned about the US negotiation strategy. In 1996, only one year before the critical Kyoto conference, the US delegation was signalling more strongly than ever before that it was willing to negotiate mandatory targets for emission reductions. To be sure, the new US position had been in the making for some time. Already in 1994, Deputy Assistant Secretary of State for Environment and Development, Rafe Pomerance, declared that the existing commitments under the UNFCCC were not enough to meet the climate change challenge (International Environment Reporter, 1994b). US industry groups were quick to react to this apparent softening of the US position and warned in a letter sent by the Global Climate Coalition to the White House that the US position as explained by Pomerance 'may be perceived as the first step towards mandatory targets and timetables' (International Environment Reporter, 1994c). With the next IPCC science assessment due in 1995, industry groups urged the US and other countries not to enter into debates on strengthening the framework convention until greater clarity on the science of global warming was achieved. In a 1994 report titled *Issues and Options: Potential Global Climate Change*, the GCC reiterated its view that the scientific uncertainties surrounding the issue meant that emission targets and carbon taxes were 'not now justified' (International Environment Reporter, 1994e). The GCC continued to maintain this position even after the IPCC's Second Assessment was published in 1995.

For US negotiators, however, the IPCC report provided justification to consider more drastic measures to curb emissions. International pressure was growing on the world's largest emitter of greenhouse gases, and the change in some business attitudes signalled that the signs were set for some sort of international agreement that would include binding international commitments. The Clinton Administration was keen to overcome the diplomatic isolation on international environmental issues that had resulted from its predecessor's refusal to sign the Biodiversity Convention and opposition to binding climate change targets at UNCED in 1992. With the State Department proposing to develop 'a progressive U.S. position' for new commitments under the UNFCCC (International Environment Reporter, 1994f), US negotiators were carefully laying the ground for a more proactive US role while trying not to antagonize corporate interests. That this would prove exceedingly difficult became clear from various initiatives with which the Clinton Administration sought to build a broader domestic alliance of environmentalists and business representatives in favour of climate action. One of the key elements of this strategy was the strengthening of Corporate Average Fuel Economy (CAFE) standards that would contribute to a reduction in GHG emissions from vehicles, which account for around 20 per cent of total US carbon emissions. In order to achieve this, the Administration formed a consultative group of environmentalists, industrialists and trade unions, the Personal Motor Vehicle Greenhouse Gas Reductions Advisory Committee. The group, however, was unable to reach a consensus on raising vehicle emission standards, and the representatives of the car and oil industries as well as trade unions refused to sign its 1996 report that called for an increase in CAFE standards (International Environment Reporter, 1996a).

Faced with a more proactive White House, US fossil fuel firms lambasted the Administration for ignoring the economic costs of its new international strategy. Their main focus was now on mobilizing opposition on Capitol Hill against international climate commitments. The real battle was now over whether the new negotiation position of the US could find support among US Senators. With both the Senate and House of Representatives under control by Republicans after their 1994 landslide victory, industry was hopeful that its anti-regulatory arguments would resonate with the new Republican leadership. Indeed, two influential chairmen of Senate committees dealing with international environmental issues took up industry's concerns: both Senator Jesse Helms, chairman of the Senate Foreign Relations Committee, and Senator Frank Murkowski, chairman of the Senate Committee on Energy and Natural Resources, expressed their reservations about the Administration's new international approach in letters sent to cabinet members in January 1994. The impact of mandatory emission targets on US industry and the failure to include developing countries in those measures featured most prominently among the concerns that the Senators expressed. In subsequent hearings on Capitol Hill, leading Congressmen such as John Dingell repeatedly criticized the Clinton Administration for seeking an international agreement that would undermine the competitive position of US firms. Industry representatives testifying for the GCC had made similar remarks in their statements, reinforcing the impression that the move towards binding targets was harmful to the US interest (International Environment Reporter, 1995a).

As the negotiations on a climate protocol neared their end, the US fossil fuel industry's strategy seemed to pay off, at least in the domestic scene. The US Senate passed a unanimous resolution in July 1997

(Senate Resolution 98, also known as the 'Byrd–Hagel Resolution'), in which it expressed its opposition to any international climate treaty that would cause serious harm to the US economy and that did not include specific commitments to limit GHG emissions by developing countries (International Environment Reporter, 1997d). The 95–0 vote left no doubt about the Senate's objection to any protocol without explicit obligations on countries such as China and India. It severely undercut the US delegation's efforts to agree on an international treaty that the Administration could submit to Congress for ratification. At the same time, however, the Senate Resolution strengthened the US position in Kyoto by reducing what Robert Putnam (1988) calls the win-set of US negotiators. Against the background of the severe constraints set by the Byrd–Hagel Resolution, the US delegation could now push for stronger concessions from other parties, particularly on the inclusion of flexibility instruments. But whether it could reach an international compromise that could be ratified at home was now only a remote possibility.

The Kyoto Protocol negotiations, 1997

As the negotiations moved closer to the 1997 deadline, the differences in US and European corporate attitudes came into sharper focus. To be sure, business opposition to mandatory emission reductions was strong on both sides of the Atlantic (International Environment Reporter, 1997c). But whereas the US Administration's efforts to draw on the support of more moderate business voices largely failed, European leaders were faced with a more diverse set of business interests, which significantly enlarged their room for manoeuvre. After UNFCCC was signed in 1992, most European industry leaders were operating under the expectation that governments would sooner or later create binding emission reduction targets. Surveys of major industries in leading countries such as the Netherlands suggested that the majority of company representatives agreed with the need to take international measures for climate change mitigation (International Environment Reporter, 1996b).

Despite opposition to strict emission reduction targets, major European firms were seeking to shape, not prevent, a protocol to the framework convention. One of the main concerns that industry leaders had was to ensure that any international reduction scheme would be business friendly, mainly by incorporating flexibility mechanisms and avoiding strict and inflexible targets. Thus, at a time when the European Commission was still seeking to introduce a carbon/energy tax as part of its international climate strategy, major industry groups such as the European Petroleum Industry Association advocated a carbon emissions trading system modelled after the US sulphur dioxide trading scheme, as well as joint implementation programmes (International Environment Reporter, 1995d). Whereas the Commission continued to favour strict commitments and a carbon tax over emissions trading well until the Kyoto conference in 1997 (International Environment Reporter, 1997b), European business groups were making the case for an international agreement that was built on flexible instruments.

The contrast with American business lobbying could not be starker. Whereas European firms applied discreet pressure to lobby their respective governments in an effort to gain influence in the design of the climate treaty, the American oil and coal industry waged a multi-million dollar public relations campaign against any binding carbon restrictions. In advertisements in US newspapers and on radio and television channels, the GCC questioned the science underlying the climate talks and warned of dire economic consequences should any mandatory restrictions on US emissions be adopted. Although failing to present a united business front, the fossil fuel industry managed to win the support of key industries, such as automobile makers, as well as farming and trade union representatives (Carpenter, 2001: 314–15). As in the run-up to UNCED, its campaign aimed at preventing, not influencing, an international climate accord.

From 1996 onwards, the growing divisions in the global business community became all too apparent in the international process. At the second meeting of the Ad Hoc Group on the Berlin Mandate (AGBM-2), in spring 1996, business representatives were for the first time unable to present a common position (Oberthür and Ott, 1999: 52). The more moderate industry groups had by then committed to supporting a successful outcome of the international process, which clearly put them apart from the US-led groupings representing energy producers and energy-intensive manufacturers. Even GCC representatives felt obliged by then to demonstrate their good faith in addressing climate change by presenting to the delegates of the Ad Hoc Group a report on voluntary emission reductions undertaken by the industry (International Environment Reporter, 1996d). But the fossil fuel lobby continued to question the underlying science of climate change and warned against basing regulatory decisions on the findings of the 1995 IPCC report (International Environment Reporter, 1996e). In the run-up to the second COP meeting in Geneva in July 1996, both the GCC and the Climate Council alleged that the IPCC's Second Assessment Report included unauthorized revisions to the final text that were not agreed by all IPCC members. The allegations were rebutted by Bert Bolin, the chair of the IPCC, and once again brought to the surface the deep divisions that existed between pro- and anti-regulatory factions of the business community (International Environment Reporter, 1996g). For the GCC's and Climate Council's actions were in stark contrast to the Business Council for a Sustainable Energy Future, which in June 1996 endorsed the IPCC findings and offered its support for efforts to reduce GHG emissions (International Environment Reporter, 1996f).

COP-1 had set up an Ad Hoc Group on the Berlin Mandate that convened a series of meetings in 1996, at which delegates sought to arrive at a draft text for a climate protocol that was to be agreed by the end of 1997. After the February/March 1996 meeting, the chair of the group was able to voice some optimism that an agreement was feasible by the deadline, despite continued obstinacy mainly by Middle Eastern oil-producing countries (International Environment Reporter, 1996g). Indeed, the major industrialized countries signalled a willingness to find a compromise and began to move on key issues that had held up earlier agreement on emission reductions. In 1996, the EU gave the first indication that it was willing to compromise on the inclusion of flexible instruments such as joint implementation as demanded by the United States. In a speech in Norway in February, EU Environment Commissioner Ritt Bjerregaard departed from her previous opposition to this form of international cooperation, and acknowledged that projects undertaken within a joint implementation framework should result in carbon emission credits for the country providing the investment abroad (International Environment Reporter, 1996c). Delegates from EU member states were more sceptical of such flexibility measures, but as the negotiations continued they came to accept that they might be the price to pay to ensure US consent.

Shortly before the third Conference of the Parties (COP-3) opened in Kyoto in November 1997, the US, together with Japan, Canada, Australia and New Zealand, once again urged the EU to lower its demands for emission reductions (International Environment Reporter, 1997e). Given the strong resistance to binding international commitments by key industry groups, the US delegation had to tread carefully as the climate talks entered their final phase and was keen to lower expectations. Other parties, too, were 'holding their cards firmly to their chests' (Oberthür and Ott, 1999: 77), as was to be expected in light of the high stakes involved. By the start of the COP-3 negotiations, it was still unclear whether a protocol or 'another legal instrument' would be the outcome. The negotiations proved particularly difficult not least due to the complexity of climate science, the long list of contentious elements of the proposed treaty and the high economic and political stakes involved.

A large number of observers from civil society, the business community and international organizations were present – nearly 4,000 in total. The participation of such a large number of observers in plenary and working group discussions caused some friction, particularly among developing countries (Depledge, 2005: 211). But while environmental NGOs and business groups played an active role in the negotiations and sought to influence their directions, the final phase of the talks was firmly in the hands of official delegates from a small number of key delegations: the United States, Japan, and the European Union. As is common in multilateral negotiations, lobbyists take a back seat when delegates move into closed-door sessions to hammer out the final details of a compromise.

The outcome of the Kyoto Protocol disappointed environmentalists but went beyond what many business lobbyists had argued for. It included a commitment by industrialized countries to reduce GHG emissions by an average of 5.2 per cent below 1990 levels, and within the commitment period of 2008–12. The agreement covers six types of GHG emissions and sets different reduction targets for individual countries or groups of countries: the US and the EU committed to reducing their emissions by 7 and 8 per cent respectively; Japan by 6 per cent; Russia and New Zealand were obliged to stabilize their emissions at 1990 levels, and Australia was allowed to increase its emissions. Developing countries were exempted from mandatory emission reductions. On insistence of the US, the Kyoto deal also included several flexibility instruments: joint implementation (JI), clean development mechanism (CDM), and emissions trading. Many details of these mechanisms, including the rules for reporting and verifying emission reductions and measures to ensure compliance, were left to be negotiated at future conferences of the parties.

Global climate governance after Kyoto

After Kyoto: the business divide grows

The adoption of the Kyoto Protocol in late 1997 marked an important step towards a global climate regime, but it proved to be just that: one step on the long road to effective climate governance. The significance of the agreement lay in the fact that leading industrialized countries had for the first time agreed to a formula of limiting GHG emissions. Five years after the framework convention, the international community had reached a consensus on specific targets and concrete mechanisms, and attention would now shift to the question of how these could be implemented. But it was not the decisive turning point in the long and tortured saga of international climate politics that environmentalists had hoped for. The targets were modest in light of the challenge that global warming poses. The agreement did not include any binding commitments for emission reductions by the fast-rising economic powers of the developing world, such as China and India. And opposition against the protocol continued to run high in some business sectors, casting a shadow over its future implementation. Furthermore, it was far from certain that the agreement would ever become legally binding. Countries representing more than 55 per cent of emissions in industrialized countries had to ratify the protocol before it could enter into force. With strong opposition against the protocol in key industrialized countries – especially in the United States, Australia and Russia – the fate of the Kyoto Protocol depended on continued diplomatic efforts to win the ratification battle.

Business reactions would prove to be of critical importance to the future success of the Kyoto Protocol. As subsequent events in the US showed, business opposition at the domestic level could easily derail ratification and/or implementation of the treaty. Moreover, business participation and cooperation was central to the functioning of the protocol's key mechanisms: emissions trading, the clean development mechanism, and joint implementation (on the protocol mechanisms, see Oberthür and Ott, 1999; Grubb et al., 1999). Finally, industry's technological power, i.e. its ability to direct investment and innovation, would become a decisive factor in determining the ability of states to steer the global economy into a carbon-reduced future.

In a sense, therefore, there were close parallels between the Montreal and the Kyoto Protocols. Both treaties were aimed at changing production and consumption patterns that were central to modern industrial societies; and the success of both depended on aligning corporate interests and patterns of business competition with the treaties' environmental objectives. But unlike ozone layer depletion, climate change poses far more complex problems that no single company or single industry can hope to solve through technological innovation. There are no substitutes that can fully replace fossil fuel-based energy, either in the short or medium term, particularly against the background of growing energy demand in developing countries such as China. Furthermore, reducing the economy's carbon intensity will require changes in production processes and products as well as consumptive patterns across all major industrial sectors. Technological innovation will thus be of central importance to climate action, but no single economic actor, or group of actors, possesses the same kind of technological power that DuPont and the chemical industry did in ozone politics.

Some of the first industry reactions were encouraging. Led by BP and Shell, European industry representatives in particular expressed guarded support for the treaty, although they did not expect dramatic changes in supply and demand patterns for oil and coal (Inter Press Service, 1997a). By contrast, the North American oil and coal lobby remained united in its opposition to the treaty. Lobbyists were outspoken in their criticism of the Kyoto deal, arguing that it lacked a scientific basis, imposed high economic costs and unfairly exempted the fast-growing economies of China and India (Inter Press Service, 1997b; Business Wire, 1997). They were joined by car industry executives who expressed their disappointment with the outcome, reaffirming the industry's longstanding opposition to firm timetables and exemptions for developing countries (PR Newswire, 1997). Environmental groups were equally critical of the agreement, though for obviously different reasons. They bemoaned the Kyoto Protocol's inadequacy in dealing with the global warming threat and noted that US energy and car producer interests had been the key driving force behind US obstinacy in the negotiations (Williams, 1997).

Amidst the mixed reactions to the Kyoto treaty, one trend became apparent as the international community moved towards implementation: despite the highly critical reaction of key North American industries, a growing number of companies started to redefine their corporate strategies. Even though ratification was still far off, the fact that an agreement had been reached shifted expectations regarding future carbon restrictions and made climate-related business risks more tangible. International restrictions on carbon-based energy sources were no longer a remote possibility but an increasingly realistic scenario. Given the uncertainty that this involved for long-term investment plans, particularly those of the energy sector (World Energy Council, 2007), many more businesses began to factor in the costs of climate action and demanded a stable regulatory environment for climate policy. The change in corporate attitudes was also being felt in the United States. The newly created Pew Center on Global Climate Change, which counted among its member companies Boeing, Enron, 3M and other leading firms, called for US leadership on global warming. As Eileen Claussen, the Pew Center's executive director, argued after Kyoto, 'the momentum has shifted to companies that accept the science. There is no question about it' (Houlder, 1998). Ford Motor Company, DaimlerChrysler, GM and Texaco left the Global Climate Coalition between December 1999 and February 2000, sending a further signal to policy-makers that business was no longer united in its lobbying effort (International Environment Reporter, 2000a, 2000b).

The period after COP-3 thus saw a growing business divide between the traditional opponents of the Kyoto Protocol and pro-regulatory forces. The origins of the business conflict on climate change went back much further, but it was after 1997 that major industrial sectors such as car manufacturing and electricity firms came out in support of international regulation. The extent to which this change in corporate strategies would undermine the anti-regulatory business front was far from clear, however. Regional and national differences – in the political and social environment and the nature of business–government relations – played a role here. Thus, while leading European companies slowly began to rally behind efforts to mitigate climate change, most American business leaders continued to resist the push for strict international regulations and remained divided on the scientific basis for, and economic implications of, carbon restrictions.

What explains the persistent differences in US and EU business attitudes on climate change? Several factors are at play here, and no single reason can be cited to explain the transatlantic gap. Corporate actors did not simply follow a straightforward economic rationale in calculating the costs and benefits of climate action. Instead, corporate preferences and political strategies were formed with reference to the wider organiza-tional and political environment in which corporations operate (Fligstein, 2002). In interpreting trends in climate discourses and politics, firms took into account not only the distributive effects that international climate action might have but also the norms and expectations that shaped the organizational field in which they operate. For many US firms, the Kyoto Protocol did not become an immediate reference point as soon as it was adopted in 1997, but continued to be a contested and potentially irrelevant form of international regulation that might never enter into force. Indeed, given the strong opposition in the US Senate against the treaty and the lack of domestic support for equivalent US regulation, it was highly unlikely that US businesses would be exposed to Kyoto-style GHG emission restrictions in the short to medium term. With the election in 2000 of George W. Bush, the oil and coal energy sector in fact gained privileged access to the Administration and was in a powerful position to work against climate-related policies in the US and internationally (Nesmith, 2002).

The international situation presented itself differently for European companies. Having already shifted to a more cooperative stance during the negotiations, many European industry groups offered to cooperate with governments in the implementation of the Kyoto Protocol. If anything, the political salience of climate change continued to rise in

Europe and businesses could therefore realistically expect the protocol to be ratified across Europe, or would at least see Kyoto-type measures enacted. For companies operating in the EU's Single Market, one of the concerns was therefore to create a regulatory level playing field across the EU, and if possible an internationally harmonized framework of emission curbs that included all major industrialized countries. European firms continued to reject carbon tax measures but were successful in pushing for greater reliance on flexibility instruments, including emissions trading (Automotive Environment Analyst, 1998; Boulton, 1998). Furthermore, in light of Europe's greater investment in energy efficiency and renewable energy sources, European firms were more likely than their US competitors to see international climate action as a business opportunity rather than a threat. On average, more European businesses tended to agree with the IPCC's Third Assessment Report of 2001, which argued that significant progress in GHG emission-related activities had been made and had been faster than anticipated.

International divisions persist in the COP meetings

After the adoption of the Kyoto Protocol, the challenge for policy-makers was now to fill the gaps in the agreement in order to pave the way for its ratification and implementation. While COP-3 had succeeded in defining specific emission reduction targets for countries listed in Annex 1 of the UNFCCC (industrialized countries), it left unresolved the question of how to achieve these targets. This was to be decided in subsequent COP meetings. The signatories of the Kyoto Protocol had committed to what environmentalists considered to be modest targets, but uncertainty concerning the economic costs of achieving these complicated the search for appropriate policy measures. Once the parties started the complex task of putting in place mechanisms and work programmes to achieve carbon reductions, the complexity of the policy problem became all too clear. Political conflicts – between the US and the EU, within the EU, and between industrialized and developing countries – further complicated the post-Kyoto process, as was evident at the COP-4 and COP-5 meetings, in Buenos Aires in November 1998 and in Berlin one year later. COP-4 produced an ambitious but ultimately vague work programme on the flexibility instruments and set COP-6 in 2000 as the target date by which decisions were to be taken on their operational aspects (Depledge, 1999). After what was widely perceived to be a 'faltering' meeting in Buenos Aires, COP-5 in Bonn produced an 'unexpected mood of optimism', but merely continued to clarify treaty elements in preparation for a bigger breakthrough at COP-6 (Earth Negotiations Bulletin, 1999: 14).

Expectations were high for the sixth COP meeting, held in The Hague in November 2000. Widely expected to complete the preparatory work for ratification of the agreement, COP-6 in fact collapsed amid recriminations between US and European delegates over their failure to narrow differences, especially on the inclusion of carbon sinks such as forests. Several factors accounted for the collapse of COP-6: the complexity of the issues on the table combined with polarized and ideologically charged country positions left little space for a political deal; and the imminent transition from the Clinton to the Bush Administration limited the scope for compromise on the US side, while the EU's difficulties in agreeing a common position weakened its stance in the talks (Vrolijk, 2001; Grubb and Yamin, 2001). A compromise was within reach, but in the end a suspension of the conference was the only way to prevent a complete breakdown of the climate talks.

At the time it seemed that the outcome of the Hague conference signalled a deeper malaise in international climate politics. Some feared that it offered an opportunity for opponents of the Kyoto Protocol to derail the process (Grubb and Yamin, 2001: 262). The COP meeting highlighted, once again, the severe domestic constraints that the US delegation was working under, but also brought out differences and conflicts within the European Union, which had been providing leadership on the path towards ratification of Kyoto (Reiner, 2001).

While climate diplomacy seemed to go round in circles, business reactions to COP-6 were more encouraging. A record number of business representatives had been in attendance in The Hague, and most of them expressed their dismay at the failure of the talks. A spokesperson of the ICC voiced disappointment at the outcome, calling for greater clarity, 'so that businesses know where they stand' (Max, 2000). DuPont's regulatory affairs manager Tom Jacob likewise acknowledged that global businesses are facing 'a carbon constrained future' but needed now to be able to calculate future risks, which required a stable regulatory environment (ibid.).

A striking feature of the international climate politics after Kyoto was the growing divergence between EU and US approaches. Whereas the EU took practical steps to implement the agreement and threw its weight behind efforts to seek ratification of a sufficient number of countries for the treaty to enter into force, the US failed to introduce domestic policies in line with its international commitment and became increasingly detached from the Kyoto Protocol. This growing divergence led to a near stalemate in transatlantic climate relations with the arrival of President George W. Bush in the White House and his decision in 2001 to withdraw from the Kyoto Protocol. By this time, it seemed that the obstruction-ist stance of the US fossil fuel industry had paid off. Despite failure to prevent an international climate treaty, the US oil and coal industries were able to undermine international climate efforts by mobilizing what was widely acknowledged to be America's *de facto* veto power in climate politics. In this sense, business and governmental interests were closely aligned. As Dunn argues, '[t]he diverging policy paths of North America and Europe have both shaped and been shaped by the strategies of firms headquartered within their borders' (2002: 28).

But closer analysis of post-Kyoto climate politics reveals that the Bush Administration's hard-line stance against Kyoto did not reflect overall US business interests. If anything, climate-related corporate interests became more diverse after Kyoto, and the ground started to shift in favour of US engagement with international climate action despite the Bush Administration's anti-environmental stance. Indeed, as developments after 2001 showed, the White House and Republican leaders in Congress became increasingly isolated amidst a groundswell of support for climate action among municipal, state-level and corporate actors in the United States. The relationship between oil and coal interests and the Bush Administration proved to be particularly close and provided core anti-Kyoto business interests with a privileged position among competing interest groups. But this position came under attack as soon as the combination of domestic political change, subnational environmental leadership and corporate support for climate action began to alter the climate agenda in US politics.

Europe embraces Kyoto's flexibility instruments

After the signing of the Kyoto Protocol, the EU turned its attention to designing appropriate strategies and instruments that would achieve the GHG emission reductions it had signed up to internationally. If past experience was anything to go by, domestic implementation would prove difficult. As discussed above, the European Commission had struggled in the early 1990s to gain sufficient support for some of its proposed climate policies but crucially failed to overcome resistance to its carbon/energy tax proposal. Despite being praised by economists as the most efficient way of tackling the emissions problem, a tax on carbon continued to be highly unpopular with key member states and industry lobbies in Europe. Although the Commission still favoured a tax solution, it gradually came to accept that this proved to be a dead-end street. The original tax proposal was officially withdrawn at the end of 2001 (Christiansen and

Wettestad, 2003: 6). Instead, the Commission began to take a greater interest in other instruments such as emissions trading.

The turn towards so-called flexible instruments marked a significant point of departure for European climate policy. Until the third COP meeting in 1997, the EU mostly resisted the inclusion of such instruments in the Kyoto Protocol and only agreed to them as part of a broader compromise with the United States, the main advocate of emissions trading and joint implementation. It is therefore somewhat ironic that the US should turn its back to Kyoto and the EU should become the main champion of an emissions trading scheme. Initially, the main focus was still on developing a global emissions scheme under UN auspices that would fulfil the commitment made by the parties in Article 17 of the protocol. But rule-making for the treaty's flexible mechanisms including joint implementation and the clean development mechanism proved to be difficult at international level and was postponed from COP-4 in 1998 to COP-6 in 2000. The slow pace of the international negotiations led European policy-makers to re-evaluate their approach and, in the words of officials of the European Commission's Environment Directorate-General, 'triggered a change of focus from top-down to bottom-up' (Zapfel and Vainio, 2002: 8). Instead of waiting for an international trading scheme to come into existence, EU leaders decided that they would take the lead and design their own system, in the hope that it would prove to be a stepping stone on the way to a global approach.

Changing business attitudes played an important role in this. The EU had no previous experience of emissions trading and came relatively late to this aspect of the debate. Until 1997, it was mostly researchers that considered different options for emissions trading in the European context (Zapfel and Vainio, 2002). Many advocates drew on the experience with existing US trading schemes for sulphur dioxide (SO_2) and oxides of nitrogen (NO_x) (Aulisi et al., 2005), but the first practical experience in Europe was gathered by corporate actors. In a widely noted speech in September 1998, BP's CEO John Browne announced that the oil company would establish an internal pilot scheme for emissions trading for a limited number of business units. This was to enable the company to achieve its publicly announced goal of cutting GHG emissions by 10 per cent below 1990 levels by 2010. The pilot scheme became operational in 1999 and was extended in 2000 to cover all former BP and Amoco operations, and again in 2001 to include the operations of the newly acquired companies Arco, Vastar and Burmah-Castrol (Akhurst et al., 2003: 662). Browne was able to announce in March 2002 that the company had achieved its goal, seven years ahead of schedule (Victor and House, 2006: 2100).

BP's experience proved to be highly influential in European debates, as it was the first such scheme that came into existence and signalled growing willingness among business groups to consider emissions trading as a less costly way of reducing GHG emissions. Zapfel and Vainio (2002: 8) consider it as a 'major driver' in the EU's shift in focus to a bottom-up approach to introducing the flexible instruments of the Kyoto Protocol. Indeed, business groups in Europe were gradually drawn to the idea of relying on emissions trading to meet the Kyoto Protocol commitments, as it would create an integrated EU-wide approach and possibly reduce the costs of compliance with Kyoto. But since very little knowledge existed at that time about how such a system would work out in practice (Brewer, 2005), business leaders as much as policy-makers acted under a veil of ignorance. Whether or not emissions trading would in the long run become an acceptable policy tool to business was still unclear. As Zapfel and Vainio argue, a 'factor that nurtured interest [in emissions trading] was the misconception of emissions trading as the cheap buy-out. This perception induced some business associations and companies fearful of effective climate policy to develop a strategic interest to continue the dialogue about the widely unknown and misunderstood instrument' (Zapfel and Vainio, 2002: 7).

The European Commission's shift in approach became apparent for the first time in a May 1999 communication, which included the possibility of a pilot scheme for EU-wide emissions trading (Commission of the European Communities, 1999). The communication acknowledged that the 'Kyoto Mechanisms are fundamentally different from the way the European Community and its Member States have organised their environmental policy over the last decades' (ibid: 14), but promised to set out in a green paper various options for an early adoption of such flexible instruments in the EU within a year. The Commission's 2000 'Green Paper on Greenhouse Gas Emissions Trading within the European Union' then signalled a growing commitment to pursue this option as a central component of domestic implementation, and in preparation for an international trading scheme under the Kyoto Protocol (Commission of the European Communities, 2000).

Introducing an ambitious emissions trading scheme in Europe not only promised an EU-wide, harmonized, approach to regulation and political buy-in from key industry groups. It was also recognized as important evidence that the EU possessed the capacity to act as an environmental leader in climate politics (Christiansen and Wettestad, 2003: 16–17).

Indeed, when the EU emissions trading system was eventually launched in January 2005, it was seen as confirmation that the EU's claim to international leadership was more than just rhetoric (Bennhold, 2005). While some companies with trading experience such as BP welcomed the scheme and called for it to be extended to the international level, many others remained initially concerned that it would hurt their position vis-à-vis US competitors. But fears of a competitive disadvantage were gradually subsiding (Bennhold, 2004), especially after the first round of emission rights allocation proved to be generous to industry (Gow, 2005).

The struggle for ratification

The EU's efforts to create the necessary infrastructure for the implementation of the Kyoto Protocol sent a clear signal to other parties that Europe was intent on making the climate treaty a success. At a time when the treaty's entry into force seemed questionable after the collapse of the COP-6 meeting and the formal US withdrawal from the agreement in 2001, European leadership in international climate politics provided a glimmer of hope that US obstructionism might not prevail in the end (Vogler and Bretherton, 2006). With the US unlikely to ratify the agreement, the hopes for the protocol now rested with Russia. In order for it to become binding in international law, ratification by at least 55 parties to the Convention representing at least 55 per cent of industrialized countries' carbon emissions in 1990 was required (Article 25). The first hurdle was passed in May 2002 when Iceland submitted its ratification. Russia's accession was now crucial as it alone could help meet the second requirement.

The ratification debate plunged Russia into a heated and polarized debate over the costs and benefits of taking climate action. 2002 marked the formal start of the process of ratification in Russia, and for over two years, government officials, parliamentarians and business lobbyists debated whether the country was better off avoiding the costs of compliance and benefiting from a warming climate and expanding agricultural areas, or whether it should commit to climate action and reap the benefits of being able to sell GHG emission rights. Against the background of fierce internal debates, President Putin was able to use the country's strategic role in the future of the protocol to extract international concessions, in the form of a substantial increase in its ability to use carbon sinks and a European agreement to back Russia's bid to join the World Trade Organization. In the end, Russia ratified the Kyoto Protocol in November 2004, thus clearing the way for the treaty's entry into force in February 2005 (Buchner and Dall'Olio, 2005; Karas, 2004).

That the Kyoto Protocol finally entered into force, eight years after the passage of the treaty and 13 years after the UN framework convention, was a significant step. Many observers had of course expressed doubt that this would ever happen, particularly so after the US pulled out of the accord (Levy, 2005: 97–8). Now that the GHG emission targets have become binding on all parties, perceptions among policy-makers and corporate decision-makers are likely to adjust to the new regulatory environment. Even if binding international commitments will be difficult to enforce internationally, governments are in a stronger bind domestically to act on their obligations and to introduce measures for their implementation. Likewise, corporate decision-makers have begun to factor in a greater likelihood of future carbon restrictions in their strategic calculations, whether they are affected by them in their home countries or in foreign operations (Hoffman, 2005). This is certainly the case in Europe, where the protocol's entry into force has further galvanized companies to explore measures to address climate change even where they are not mandated (Mesure, 2007). The effect could also be felt in the United States, where a number of leading corporations have moved ahead of their federal government in adopting progressive climate strategies (Hoffman, 2006).

The emerging multi-level governance of climate change

As the parties to the Kyoto Protocol now work to fill the remaining gaps in its institutional architecture and prepare for a successor agreement, the focus has shifted to the role that business actors play in climate governance more generally, within and outside the Kyoto framework. The ratification of the Kyoto Protocol was an important milestone towards effective climate policy, but action against global warming will have to be taken on a much larger scale than is prescribed in this treaty. For one, the Kyoto Protocol's emission reduction scheme will run out in 2012, and negotiations on future commitments are already under way. At the same time, climate action has shifted to other forums, located at the subnational, regional and transnational level. Global climate governance now extends well beyond the architecture of climate governance built around the UNFCCC and Kyoto Protocol, and includes a growing web of private as well as public-private institutions that provide governance functions without relying on the authority of the nation-state. Jagers and Stripple refer to this as the broader notion of global climate governance that includes 'all purposeful mechanisms and measures aimed at steering social systems towards preventing, mitigating, or adapting to the risks posed by climate change' (2003: 385). Business actors have played an active role in this emerging field of global climate governance, driven by the growing political salience of climate change, under pressure from civil society and governments, and in search for business opportunities.

A number of corporations have initiated programmes that complement national and international climate regulations, but have also sprung up outside the contexts of mandated climate action. For many companies that emit greenhouse gases, a first and critical step has been to measure their contribution to the global warming effect, and to create accounting and reporting standards that would enhance transparency. Many companies have combined these efforts with programmes to identify energy savings and/or emission reductions, and some have gone as far as teaming up with environmental organizations to agree targets and establish verifiable indicators of success. In a final step, some companies have joined forces to create markets for GHG emissions, in an effort to spearhead the development of flexible instruments. While business actors have provided the critical input for these initiatives, many have been developed among a wider network of partner organizations, involving industry, environmental consultancies, NGOs and public institutions.

Major global firms (e.g. BP, DuPont, Ontario Power Generation) together with Environmental Defense formed the Partnership for Climate Action, a grouping that established a commitment to reduce carbon emission by 80 million tonnes by 2010 and to create an emissions trading system among them. WWF has established a scheme with companies such as Johnson & Johnson and Nike, the Climate Savers Programme, to achieve greater efficiency and changes in energy sources as part of a wider climate and energy strategy. And the Pew Center on Global Climate Change has set up a Business Environmental Leadership Council in cooperation with leading multinationals such as Alcan, Deutsche Telekom, IBM and Novartis, which promotes strategies for specific emission reductions (Dunn, 2002: 34).

The Chicago Climate Exchange (CCX) is a voluntary emissions reduction and trading system, the first of its kind in North America. It opened for business in early 2003 and has seen membership soar to 210 by mid-2006. Its 14 founding members, which include energy (American Electric Power), chemical (DuPont) and automobile (Ford Motor Co.) firms, as well as municipalities (City of Chicago), are committed to reducing GHG emissions by 2 per cent below 1999 levels during 2002 and reducing them by 1 per cent annually thereafter. The CCX has attracted growing corporate support as it provides the only framework in which to develop and test GHG emissions trading as a tool for climate governance (Dunn, 2005: 38–9; Grant, 2004).

Corporate climate strategies have also been institutionalized at the international level. The Carbon Disclosure Project (CDP) is one of the new initiatives that seek to enhance the informational environment of global climate governance. Launched in the UK in 2000, it brings together institutional investors such as pension funds in an effort to create greater transparency on the role that publicly listed corporations play in causing and mitigating GHG emissions. This information is gathered with a view to informing investors about how companies and their long-term economic position may be affected by climate change and future climate policies, especially restrictions on carbon emissions. Starting in 2002, the CDP has issued annual requests for information, initially from the 500 largest corporations as listed on the FTSE 500 index, rising to 1,800 listed companies in 2006. At the same time, the investors represented by the CDP have risen between the same period from 35 to 211, with the assets that they are managing rising from a total of over $4.5 trillion to over $31 trillion. The CDP is best viewed as an auxiliary governance mechanism. By gathering and disseminating investment- and policy-relevant information about the exposure of large companies to climate risks, it enables other actors and governance mechanisms to direct investment in a more climate-friendly direction. Its effectiveness thus depends not only on the quality of the information it provides but also the way in which it is used by others in the pursuit of commercial or environmental objectives (Daneshkhu and Harvey, 2006; <http://www.cdproject.net>).

These corporate initiatives have assumed a more direct political relevance in the context of American politics, where businesses together with political leaders from states and municipalities have challenged the Bush Administration's refusal to adopt GHG emission reductions. In a sense, the absence of the United States from the Kyoto Protocol has created a policy vacuum at the national level that other actors at the subnational level have rushed to fill.

One of the earliest subnational initiatives arose in the context of the Cities for Climate Protection (CCP) programme, an international coalition of municipalities that have pledged to address the threat of global warming through changes in their policies on land use, transportation, energy management and planning. The CCP programme was launched in 1993 by the International Council for Local Environmental Initiatives and counted 674 members worldwide in 2006, which accounted for around 15 per cent of global GHG emissions. US municipalities have played an active role in this programme and have initiated their own national campaign within the CCP. In 2005, the US Conference of Mayors went one step further. It declared its intention of meeting or exceeding the Kyoto Protocol's emission reduction target for the United States and called on state and federal governments to do the same (Betsill and Bulkeley, 2004; Selin and VanDeveer, 2007).

Several US states have also developed proactive policies to address climate change. Given the political inertia at the federal level, their actions have assumed political significance both nationally and internationally (Rabe 2004). According to the Pew Center on Global Climate Change, 30 states have joined regional climate change initiatives and 39 states have created GHG inventories.[1] Several governors have set emission reduction targets and a few leading states have enacted laws that provide a basis for implementation. In a landmark deal that set the scene for a more aggressive approach at the state level, California's governor and state legislature agreed in 2006 to achieve a 25 per cent reduction in GHG emissions by 2020 and introduce emission controls that apply to utilities, refineries and manufacturing plants (Barringer, 2006). California has since joined four other states in the western US to create a regional effort to reduce emissions (Martin, 2007). Similar collaborative schemes have come into existence in the north-east, including the New England Climate Coalition and the Regional Greenhouse Gas Initiative (Selin and VanDeveer, 2007: 4–6).

These state level and municipal initiatives have had two effects on business perceptions and strategy. The growing diversity of climate policy in the United States has increased concerns among corporations that they will have to operate in a more fragmented and uncertain regulatory environment. At the same time, the growing support for subnational climate action has raised expectations that the political gridlock over climate policy at federal level might soon be broken. Especially after the Congressional elections of November 2006, which handed control of both the House and the Senate to the Democratic Party, many corporate leaders have concluded that the US is headed for mandatory carbon restrictions at federal level. The changes in the political landscape have created a powerful rationale for hitherto reluctant business actors to embrace the idea of more progressive climate policies, as has become evident in recent corporate testimonies on Capitol Hill (Donnelly, 2007).

Will the change in business strategy that has become apparent across major US industries directly translate into political change in the US, and thereby strengthen the international climate regime? The business sector has undoubtedly played a powerful role in shaping America's climate policy and helped to prevent US participation in the Kyoto Protocol. Now that the business sector has grown more divided and the fossil fuel industry's influence has declined, should we expect an early and decisive shift in US policy? At first sight, the business conflict model would suggest that growing divisions among previously united business actors open up political space for new political coalitions in favour of policy change. But it would be a case of misplaced economic determinism to argue that this outcome is inevitable. While many US business leaders have started to lend their support to mandatory emission reductions, others remain sceptical. The ground has shifted, but the balance of competing business interests remains uncertain. Furthermore, the Bush Administration continues to oppose such measures, and even if a new Administration were to adopt a more proactive climate strategy, it remains unlikely that it could gain support in Congress for ratification of the Kyoto Protocol. Business conflict has thus opened political space, but in a neo-pluralist context, the future direction of US climate policy remains undecided and subject to shifting political alliances and discourses. Some observers point out that a political window of opportunity for fundamental change now exists (Selin and VanDeveer, 2007), but whether it will suffice to re-engage the US in the international climate regime remains uncertain.

Still, the overall significance of the changes in corporate strategy is clear. At a discursive level, it has helped to move the debate from whether there is sufficient scientific evidence for manmade global warming to the question of how societies and industries might best respond to the climate challenge. Of course, doubters of climate science are still vociferous, especially in some US business circles, but a growing field of business leaders have come to accept the need for precautionary action irrespective of the remaining scientific uncertainties. More and more businesses are exploring ways of positioning themselves as climate leaders in their sectors, hoping to gain a first-mover advantage or seeking to create synergies between climate action and other corporate strategies (Hoffman, 2006; Cogan, 2006). Whether these corporate activities can have a significant impact remains to be seen, though the discursive shift they have promoted is in itself noteworthy. The World Energy Council recently captured this new business sentiment in a policy statement of March 2007, in which it stated that leading electricity companies agree that 'addressing climate change now will be less risky and costly to the world economy than postponing action', and that '[t]aking bold, early steps to curb greenhouse gas emissions appears to be profitable for business, government and consumers' (World Energy Council, 2007: 1).

Corporate initiatives that create greater transparency about their contribution to climate change, establish mechanisms for GHG reductions and introduce climate-friendly technologies have all contributed to a further discursive shift that plays directly into international policy-making. Collectively, these initiatives have confounded earlier warnings by business lobbyists that reducing climate-damaging emissions would be technically impossible or too expensive, and have therefore shifted the regulatory debate further in the direction of precautionary action. For international climate policy has been held back not only by the scientific uncertainty but also by the perceived technological barriers and economic costs of climate measures. This form of technological uncertainty has been one of the constraining factors in developing an international climate regime, as was the case in ozone layer depletion. It directly affects the calculations of interest by states and business actors, and the growing support among corporations for climate action has therefore helped to reduce perceptions of cost and technological uncertainty. Recent model calculations, such as the UK's Stern Review on the economics of climate change (Stern, 2007), have also supported a downward adjustment of the cost estimates of climate mitigation.

Finally, the recent climate strategies of leading firms have opened new opportunities for environmental campaign groups to engage business in cooperative endeavours to promote and institutionalize climate action. The multilevel climate governance structure that has emerged in recent years includes nonstate mechanisms in which social and business power meet to create global governance functions. In this way, the diversification of business strategies has helped to empower NGOs and other actors, including international organizations.

Conclusions

Few would dispute that business has had a powerful influence on the international politics of climate change. Fossil fuel industry groups have been present in international debates from an early phase, and as the international effort to create a climate regime gathered pace, a vast number of business representatives attended the international negotiations and lobbied governmental representatives. Business has been in a powerful position for structural reasons, too. Virtually every major industrial sector contributes to global warming, and therefore also plays a key role in the search for technological solutions. This has given business a high degree of structural power, allowing it to slow down international regulatory efforts, but potentially also to help in addressing the problem. There is no single sector, however, which holds the keys to resolving this complex environmental threat, not even the fossil fuel energy sector. Climate change and its causes reach into every corner of the modern industrial system, and the changes required to patterns of production and consumption are so extensive that no single business sector can be seen to 'control' international climate politics. It is therefore the complexity of climate change as industrialism's major ecological shadow that stands in the way of timely and drastic measures to combat global warming.

The above analysis has nevertheless shown how business actors have shaped international climate policy in important ways. During the early phase of international climate politics, the industrial sectors most directly threatened by the creation of international restrictions on GHG emission were the first to organize themselves, first in the United States and then internationally. Through organizations such as the Global Climate Coalition, leading oil and coal companies together with major industrial manufacturing sectors created one of the most effective lobbying groups that dominated business representation at the international level well into the 1990s. Led by the oil multinationals, the fossil fuel industry became the dominant business lobby group in the 1992 'Earth Summit' negotiations on the framework convention and in the run-up to the 1997 Kyoto Protocol talks. During this phase, competing business interests that were more open to international regulation also began to involve themselves, though with limited success. The renewable energy firms were economically weaker and more fragmented as an industry, especially when compared to the oil industry, and the insurance industry failed to develop an effective political strategy to shape the international process.

Business conflict was therefore always a possibility in climate politics but did not become a more powerful pattern in business lobbying until well into the mid-1990s. Of more significance was therefore the growing disintegration of the fossil fuel industry's lobbying campaign, as a growing number of manufacturing industries distanced themselves from the oil industry's hard-line opposition to climate action and took a more conciliatory stance before and after the Kyoto conference of 1997. Differences in strategy also emerged within the oil sector, especially between US and European firms, and this further promoted the emergence of greater business conflict and a more diverse field of corporate climate strategies. New groupings such as the International Climate Change Partnership and the World Business Council for Sustainable Development were now openly supporting international climate action, as long as it provided for flexible and market-friendly instruments.

Unlike in ozone politics, such shifts in business sectors did not translate more directly into shifts in the negotiation dynamic, although they shaped the discursive environment in which more progressive climate policies have been sought. The US government followed most closely the predominantly anti-regulatory stance of its fossil fuel industry, while European governments were able to push for more stringent GHG restrictions as European business leaders pledged their support for climate action. Business conflict thus helped to reinforce the growing shift towards acceptance of climate science and of the need for precautionary action. Particularly since the adoption of the Kyoto Protocol, the growing diversity of business interests and strategies has created new opportunities for progressive political alliances in favour of climate action and governance, at the sub-, inter- and transnational level. Business support for climate action has always been conditional on the adoption of flexibility instruments within the Kyoto Protocol and a regulatory environment that favours market-friendly instruments and targets and timetables that reflect the pace of technological change in industry. This is one area where business actors have been particularly influential in the design of the emerging climate regime.

Overall, business has been unable to control the emergence of the global climate agenda but has significantly slowed down the rate at which international regulations have been adopted. It has shaped the regulatory discourse, reluctantly confirming the need for precautionary action while pushing for flexible and market-friendly instruments. Business conflict has weakened the original anti-regulatory stance of the fossil fuel industry and has opened up political space for states and NGOs to push for stricter international measures. But ultimately, it is changes in the underlying structural importance of fossil fuel-based energy that will determine the overall impact of business conflict in the politics of climate change.

5
Agricultural Biotechnology

Genetic engineering is reshaping agricultural trade and food production worldwide. The development and commercialization of genetically modified (GM) crops is the fastest technological revolution that has ever occurred in agriculture. In less than two decades, genetically modified organisms (GMOs)[1] have moved from laboratory research through field testing to commercial production, with the global GM planting area growing at around 10 per cent annually in the last decade. Led by Monsanto, a small number of powerful biotechnology firms have set out to reshape global agricultural markets for key internationally traded crops such as soybean, corn and cotton. More GM crops (e.g. rice, wheat, potatoes) are in the pipeline.

But the progress of agricultural biotechnology has not been straightforward. Resistance to GM food production has sprung up in various places, among consumers, food producers, retailers, farmers, and regulatory authorities. European consumers widely rejected GM food when it became available for the first time in the mid-1990s, and opposition has also emerged in Latin America, Africa and Asia. Many developing countries are still weighing the pros and cons of adopting GM technology in their agricultural systems, with some African countries even rejecting GM food aid. Agricultural biotechnology became one of the *bêtes noires* of the anti-globalization movement of the late 1990s, and continues to ignite fierce public debates on how to reduce environmental risks, ensure food safety and control the power of multinational corporations. In Europe, a system of precautionary GMO regulation has been put in place to test and monitor the environmental and health risks of GMOs. Other nations have followed suit and have created their own safety regulations and laws. And at the international level, a binding treaty on safety in GMO trade has been established, despite lobbying by business groups as well as opposition from the United States, Canada and other pro-GMO countries.

This chapter examines the evolution of international biosafety politics, from the first international debates on biosafety in the 1980s to the negotiations on the Cartagena Protocol on Biosafety in the 1990s and beyond.

The science and business of agricultural biotechnology

Genetic engineering in agriculture

The manipulation of genes in living organisms has opened up a vast range of commercial applications, from medicine to agriculture and environmental remediation. The genetic modification of agricultural crops became a commercially viable prospect in the 1980s, when biotech researchers were experimenting with different uses of recombinant DNA techniques in plants and animals. Genetic engineering has been used to insert desirable traits into living organisms or to make plants resistant to pests, drought, herbicides or other environmental stresses. Several different GM plants have since been developed, most notably soybeans, cotton, corn, canola and rice. The global market for GM crops is valued today at over $6 billion (Davoudi, 2006), and is second only to the medical uses of modern biotechnology.

Scientists, environmentalists and consumers have expressed concerns about the commercial introduction of agricultural biotechnology. Critics of the technology point out that despite decades of research, too little is known about the potential long-term effects of GMOs on ecosystems and the human body, and that policy-making should therefore be guided by precaution. With regard to the environmental effects of GMOs, scientists have raised concerns that genes from plants that are modified to be herbicide-resistant could accidentally be transferred to other plants, thereby spreading herbicide-resistance with undesirable results; that farmers would use greater amounts of herbicides without fear of crop damage in cases where herbicide resistance is created in GM crops; that insect-resistant crops would have a detrimental effect on non-target insect populations, and by implication bird populations that feed on insects; and that insect-resistant crops might lead to resistance by target insects to the toxins contained in the GM crops. Human health concerns over the introduction of GM ingredients into the food chain centre on toxins contained in GM crops that pose a threat to humans; proteins introduced through genetic engineering that may cause allergenicity; and antibiotic marker genes used in the genetic modification of crops, which may be transferred to the human body causing resistance to commonly prescribed antibiotics. These concerns have informed the emergence of a transnational anti-GMO movement, although it is important to note that it comprises a wide range of views, from more radical voices that reject genetic engineering *per se* to reformist environmentalists who advocate a case-by-case risk assessment (Bauer et al., 2002).

As in many other environmental areas, the scientific community is far from united in its assessment of the environmental and health risks of GM crops. Many years of research and field trials have so far failed to produce a conclusive answer as to whether or not the new technology is safe. A growing number of GM crop varieties have been approved for planting and human consumption, particularly in North America, but no global mechanism exists to reconcile the existing differences in risk assessment and regulatory approval. The situation is further complicated by the long-term perspective that ecological risk assessments demand, which in the case of biodiversity impacts can stretch to several decades. Environmentalists have therefore called for the application of the precautionary principle in risk assessment and management, which has also emerged as a guiding norm in international environmental law (Birnie and Boyle, 2002: 115–21).

The global production chain of GM food

To understand the conflicting dynamics that have shaped the evolution of agricultural biotechnology and the creation of a global biosafety regime, it is important to consider the different corporate actors that are part of the global production network of GM food. The chain contains a large number of corporate players from the biotechnology, farming, commodities trade, food production and retailing sectors. Viewed from the consumer end of the chain, the GM food chain appears to be a buyer-driven production network, where large food retailers exercise considerable influence over the sourcing and production of food products. Such buyer-driven food networks are usually nationally or regionally organized, based on national or regional patterns of retailing and consumption. At the same time, however, we can also view the GM food chain from the producer end, where producers and distributors of GM seeds have undergone a process of industrial concentration. Indeed, the producer end of the chain has seen a trend to create a producer-driven network controlled by biotechnology firms, with agro-chemical and biotechnology companies developing an integrated model of seed production and marketing.

The barrier to further integration of the GM food chain from both ends has been the highly fragmented farming sector and the diversity of national and regional agricultural markets. Neither food retailers nor seed producers have managed to establish full control over the global food chain, although recent processes of economic concentration at both ends have significantly increased the influence of large producers and buyers. What is interesting to note is that GM food is part of a long production chain that is transnationally organized and that contains a large number of different types of corporate actors with diverse interest structures. This in turn offers multiple entry points for social and political actors to seek to influence the production of and trade in GM food products and to form political alliances with corporate interests that are opposed to the strategies of the biotechnology industry (Schurman, 2004).

The field of agricultural biotechnology was developed initially by a large number of small and medium-sized companies, many of which had started out as spin-offs from research institutes and universities. To bring new biotech products to the market required substantial funding sources and involved sizeable commercial risks. Governmental support for research and venture capital investments in the commercialization of new products helped the emerging industry to grow. Still, many small firms faced severe financial constraints and were gradually taken over by the life sciences industry that had been built around pharmaceutical applications. By the 1990s, the highly fragmented nature of the early biotechnology sector gave way to a more consolidated industry with a small number of large players, most of them located in North America, Europe and Japan.

US firms have led the commercialization of biotechnology from early on and have since been able to cement their leadership position. Europe's biotechnology firms started out later, have been smaller on average than their US competitors, and have been spending less on research and development. Public funding for biotechnology has been more generous in the US than in Europe, too, and much of European biotechnology investment has been channelled into US-based research institutes. Indeed, in order to tap into the stronger technology and human capital base of the US, leading European firms such as BASF, Ciba-Geigy and ICI have expanded their US presence by setting up US research sites, thereby further entrenching the global dominance of US biotechnology (Kathuri et al., 1992; Dibner et al., 1992).

As GMO innovations were moving from laboratory tests to field trials and commercialization, the emerging biotechnology sector underwent a process of industrial consolidation. From the mid-1990s onwards, a wave of mergers and acquisitions has paved the way for a radically different industrial landscape that saw a handful of large biotechnology firms dominate GM crop development. The UK's Astra and Sweden's Zeneca, two large pharmaceutical firms with stakes in agri-biotechnology, merged in December 1998 to form the new company AstraZeneca. Only a year later, AstraZeneca and the Swiss pharmaceutical producer Novartis decided to spin off their respective agro-chemical businesses and merge them to form Syngenta. And in April 2000, Monsanto and Pharmacia & Upjohn completed a merger of their pharmaceutical production and the creation of a separate company focused solely on agri-biotechnology, under the name of Monsanto (Fulton and Giannakas, 2001).

What was behind this wave of mergers was the desire to achieve synergies particularly in the pharmaceutical sector and to broaden the application of genetic engineering techniques to other areas. But the creation of Syngenta and Monsanto, with a sole focus on agri-chemical and agri-biotechnological business, also suggested that the agricultural and the much larger medical sectors were increasingly going their own ways. Against the background of a worsening public climate for GM crops and regulatory restrictions particularly in Europe, integrated medical and agricultural biotech firms were keen to separate out the different social, political and economic risks involved in biotechnology and to shield medical applications from the negative publicity that GM food attracted (King et al., 2002).

A different motivation lay behind the second wave of mergers that saw DuPont acquire Pioneer Hi-Breed in 1997, to become the world's largest seed company. Monsanto had kicked off this wave when it decided to take over DeKalb in 1996, and followed this up with further acquisitions in the late 1990s, including Holdens, Delta & Pine Land Co., Asgrow and Agracetus (Joly and Lemarié, 1998). Both DuPont and Monsanto pursued these acquisitions as part of a broader strategy to integrate crop development, agrochemicals production and seed distribution. This would give them greater control over the entire seed and agrochemicals business and would put them in a strong commercial position as the sole suppliers to farmers in certain market segments. Moreover, biotechnology firms saw the creation of value-added crops as leading to the emergence of specialty markets with premium prices (Shimoda, 1994). While the industry is still far away from this vision, it has achieved near-oligopolistic, and in some cases monopolistic, control over the supply of seeds such as GM soybeans and cotton in North America. This has become a key source of power for biotech firms, vis-à-vis farmers and regulators.

Today, after a process of industry consolidation, less than a handful of companies control the global market for GM crop varieties, with Monsanto being by far the dominant player. In 2005, GM crops were grown on an estimated 222 million acres around the world, of which Monsanto's GM crops were grown on 217.2 million acres, more than 90 per cent of the total biotech acreage. The next largest biotech companies, Dow/DuPont, Bayer and Syngenta, only accounted for 4.7, 6.1 and 7.9 million acres respectively. Other firms were responsible for only a further 1.8 million acres worldwide. Thus, four US and European firms dominate the global market for biotech products, and out of these Monsanto remains the undisputed market leader by a long margin (Davoudi, 2006).

The global GM crop area has grown steadily for the last ten years, at an average annual rate of around 10 per cent. Still, only five countries produce the vast majority of all GM crops: the United States, Argentina, Brazil, Canada and China. The US alone accounts for over half of the world's GM crop production. Despite a decade of year-on-year growth of GM crop planting, the biotech revolution has so far failed to spread worldwide. Many of the large agricultural markets, such as the European Union, Japan and Korea, have put in place stringent import regulations, including GMO labelling requirements and partial or outright bans on GMO imports, and in many countries where certain GM crops have been authorized for commercial sale, consumers and food retailers are refusing to buy or stock GM food products.

Creating a global biosafety agenda

Genetic engineering and the emerging biosafety agenda

The first safety debate on genetic engineering arose in the United States, the home of the world's leading biotech research institutes. In the early 1970s, scientists raised concerns about laboratory experiments with GMOs. After a one-year voluntary moratorium on such experiments, the safety issue was debated at a major international conference at the Asilomar Conference Center in Pacific Grove, California, in February 1975. It produced a broad consensus among scientists that experiments with genetic engineering, while potentially hazardous, could be controlled with adequate containment measures. The conference concluded that most research could proceed under a voluntary code of conduct and recommended the lifting of the moratorium that had been in place since July 1974 (Wright, 1994: 136–57; see also Krimsky, 1982: chapters 7–10).

The Asilomar Conference had an important impact on political responses to GMO safety concerns in the 1970s, largely assuaging the fears that GMOs might turn out to be biohazards. One of its most important results was that it reaffirmed most governments' belief in a system of scientific self-regulation. In the US, the National Institutes of Health (NIH) adopted laboratory safety guidelines in 1976 that became a blueprint for other countries' safety guidelines and confirmed the broad consensus on a 'light touch' regulatory approach (Wright, 1994: 157–63).

At this time, most genetic engineering research was conducted in research laboratories without much involvement by the corporate sector, and commercial applications were still far off. This, however, began to change in the 1980s, when genetic engineering moved from the laboratory to field trials. Accordingly, the safety debate shifted from the physical and biological containment of GMO experiments to the ecological effect of GMO releases into the environment (Walgate, 1990: 168–9). Slowly but steadily, the pressure on governments to regulate modern biotechnology grew. In the United States, Congress started to discuss how to deal with the growing regulatory uncertainty that arose in the early 1980s, when the first field trials of GMO were being authorized by NIH committees. The US Environmental Protection Agency (EPA) sought to gain regulatory authority for biotechnology, citing the potential threat that GMO releases posed for the environment. However, anti-regulatory forces within the Reagan Administration quickly sought to prevent such a development and convened a cabinet-level working group on biotechnology to resolve the question of regulatory authority, thus signalling a more business-friendly approach to biotechnology (Sheingate, 2006: 247–8). The resulting 'Coordinated Framework for the Regulation of Biotechnology', which was adopted in 1986, created a divided system of regulatory authority, shared between the EPA, the Food and Drug Administration and the US Department for Agriculture.

The 1986 'Coordinated Framework' had important ramifications for the future of biotechnology in the US. It was clear that the more pro-business-oriented White House had prevailed and was able to prioritize the economic potential of biotechnology over environmental and health concerns. US regulators adopted a product-based approach, which meant that biotechnological products were to be treated differently from other plants or food products only if they exhibited certain traits or characteristics that were considered to be potentially hazardous. Rather than applying a technology-based regulatory framework that would apply to all biotech products because of the process that was involved in their manufacture, US authorities assumed substantial equivalence between biotech and conventional products. The Reagan Administration was keen to pave the way for what it saw as a promising new industrial sector and did not want environmental fears to stand in the way of this vision.

Most European governments were equally keen to promote the future industrial potential of biotechnology. When the US adopted the 'Coordinated Framework', it looked as if Europe was to follow the same regulatory path. However, developments towards the end of the 1980s would create a gulf between the two major biotechnology regions that was to grow deeper in the 1990s.

During the mid-1980s, the EU was only in the process of developing an environmental policy competence and had no authority to regulate the environmental risks of genetic engineering. Instead, EU member states developed their own policy responses, which resulted in an uneven regulatory environment. Denmark and Germany, two countries with a strong domestic environmental movement, were the first to introduce comprehensive gene laws, in 1986 and 1990 respectively, setting a precedent for comprehensive, precautionary, process-based regulation. Other countries, including the UK and France, continued to rely on limited safety regulations involving voluntary or mandatory notification requirements, while some (e.g. Italy and Luxembourg) did not create any GMO-specific regulations. With the exception of Denmark, and to some extent Germany, EU member states shied away from drastically increasing regulatory oversight. As a survey of the regulatory landscape in the mid-1980s commented, the situation in Europe was 'characterized by flexibility and relatively minimal constraints, on the basis of the traditional freedom of scientific inquiry' (Mantegazzini, 1986: 82). Throughout this period, the biotechnology sector enjoyed strong political support in European capitals and in Brussels, where the European Commission's Science Directorate-General championed its cause (Patterson, 2000: 321–3).

As the first commercial applications of genetic engineering appeared on the horizon in the 1980s, the commercial promise of the emerging biotech industry became a key focus in public policy on both sides of the Atlantic. The US, and to a lesser extent Japan, were clearly in the lead in both basic and applied biotechnological research, causing a considerable amount of soul searching and a flurry of new policy initiatives in European capitals (Paugh and Lafrance, 1997: 99–102). This was also the time when the European Union went through one of its most dramatic phases of economic and political integration, with the signing of the Single European Act in 1986 and the launch of the Single European Market programme that was to be completed in 1992. The perceived gap between American and European biotechnology played into the hands of those EU officials who were keen to create a policy competence at the European level. Regulatory harmonization via the Single Market project thus became a key instrument for creating a more unified European approach to biosafety regulation (Mantegazzini, 1986: 99; Patterson, 2000: 320). But in a state of regulatory uncertainty and political change, the pro-environmental forces of the newly created EU Environment Directorate-General managed to wrest regulatory powers from competing DGs. That this would result in a regulatory environment that was to emphasize environmental regulation over industrial promotion is one of the unintended consequences of the EU's harmonization drive (Hodgson, 1992).

The EU established a European-wide system of biosafety regulation in 1990, in the form of two Directives on the contained use of GMOs (Directive 90/219/EC) and on GMO releases into the environment (Directive 90/220/EC). The 1990 regulations established for the first time a horizontal, process-oriented path to regulating biotechnology in Europe, in sharp contrast to the more limited approach in the United States that presumed substantial equivalence between biotech and conventional products. European regulators also introduced the precautionary principle, allowing authorities to prevent GMO releases and commercialization under conditions of scientific uncertainty, i.e. without proof of harm. This was also to have profound consequences for the EU's later role in the international biosafety negotiations during the late 1990s, laying the foundation for the EU's environmental leadership role (see Falkner, 2007b).

Much to the frustration of the nascent biotech industry, regulatory harmonization thus reversed the previous trend towards a business-friendly biotechnology policy in Europe as the biotech sector was only weakly involved in the drafting process. European biotechnology companies were slow to form EU-level associations and to lobby European institutions at a time when the regulatory environment was in flux. They were held back by a tradition of organizing around products, not industrial processes, and a fragmentation of the biotechnology sector into small and medium-sized firms (Greenwood and Ronit, 1995). It was only in June 1989, when the EU's legislative drafting process was nearing its end, that the Council of the European Chemical Industry created the Senior Advisory Group for Biotechnology (SAGB), the first European industry group on biotechnology comprising leading agrochemical firms such as ICI, Sandoz, Rhone-Poulenc, Hoechst and Monsanto Europe (Hodgson, 1990b; Cantley, 1995: 633–4). By that time, industry observers were beginning to suggest that the regulatory battle in Europe may have already been lost (Hodgson, 1990a).

By the early 1990s, sharp differences in the developmental paths of biotechnology between North America and Europe had appeared.

US biotech firms were leading the global race to develop commercial applications of genetic engineering. They operated within a favourable regulatory environment and benefited from a vibrant venture capital sector that was willing to invest in high-risk enterprises spun off from smaller research institutions. By comparison, European biotechnology companies were smaller in size, less well funded and poorly integrated across Europe. Crucially, they were operating in an increasingly hostile regulatory and social environment, with anti-GMO sentiment running high in several EU countries. The late 1990 EU directives were seen as the beginning of a slippery slope by biotechnology firms as they introduced the principles of process-based and precautionary regulation. This divergence of industrial and regulatory paths was to have consequences for the economic and political strength of biotechnology firms on both sides of the Atlantic. Whereas the US biotechnology industry developed into a powerful industrial sector that built up a robust PR and lobbying machinery in the form of the Biotechnology Industry Organization (BIO), European companies' attempts to counter rising anti-GMO sentiment through organizations such as EuropaBio remained low key and ineffective. As Willy De Greef, head of regulatory and government affairs for the Swiss multinational Novartis Seeds, commented in 1999, 'BIO is a fantastic machine, but Europabio can never be that powerful because the subscriptions feeding it are not of the same size as in the US because the business is not of the same size' (Dorey, 1999: 631).

1992: Biosafety at UNCED and in the Convention on Biological Diversity

Just how large the gap had grown between the US and EU biotech industries became apparent during the preparations for the 1992 UN Conference on Environment and Development (UNCED). The negotiations on the UNCED programme of action ('Agenda 21') and the Convention on Biological Diversity (CBD) provided the first opportunity to create international rules for agricultural biotechnology. They also produced the first showdown between pro- and anti-regulatory forces that was to be repeated again in the international biosafety politics of the 1990s. The US biotechnology industry followed these negotiations and strongly opposed the creation of binding international rules on the safety of biotechnological processes and products, or what became known as 'biosafety'. They were supported in their stance by the two leading biotechnology countries, the United States and Japan. By comparison, European biotechnology firms played only a marginal role in the negotiations and were less concerned about the implications that the CBD might have for their commercial interests. Whereas the US government emerged as the most uncompromising defender of the biotechnology industry and eventually refused to sign the Biodiversity Convention out of fear that it might harm industry, European government struck a more conciliatory tone and signed up to the CBD.

Developing countries were the key *demandeurs* for binding international biosafety rules. Their proposals in the UNCED preparatory meetings reflected a number of concerns: that potentially dangerous biotechnology products could be released into the environment without adequate safety provisions; that northern biotechnology firms might move some of their operations to the developing world without sufficient regulatory controls; and that the use of GM crops might harm established agricultural systems and farming communities (Zedan, 1992). Many developing countries were suspicious of what they perceived to be a Northern technology and drew parallels with the threat from international trade in hazardous waste, which was addressed in the 1989 Basel Convention (Clapp, 2001).

At a time when genetic engineering had yet to reach the stage of commercialization in agriculture, leading biotech countries saw such regulatory proposals as unnecessary and potentially harmful to their economic interests. The US left no doubt about its preference for domestic over international policy-making and sought to eliminate all references to the safety of biotechnology from the draft text of Agenda 21 (Munson, 1993: 499). US delegates were particularly opposed to some of the more far-reaching demands by countries such as Malaysia, which entailed the creation of legally binding safety measures and a system of liability and redress. The Bush Administration had unmistakably stated its deregulatory philosophy in February 1991, when the President's Council on Competitiveness issued a report on national biotechnology policy that industry observers praised for being 'very favourable indeed to commercial applications of the biotechnologies' (Bio/Technology, 1991; see also Fox, 1991). It was not going to allow the UN to reverse the trend towards a more investor- and industry-friendly climate for biotechnology. Most European countries conceded that an international instrument on biosafety, in the form of non-binding safety guidelines, was desirable, but Europe's leading biotech countries also opposed binding international rules. In the end, the EU supported a compromise agreement on Agenda 21 that included a reference to internationally agreed guidelines on safety in biotechnology releases (Agenda 21, chapter 16).

The other arena in which biotechnology regulation became a controversial issue was the CBD. On most though not all issues, the negotiations again pitted developed against developing countries. The former were keen to see environmental protection measures as the cornerstone of the convention, while the latter pushed for an agreement focused on North–South equity issues, particularly access to genetic resources. From the 1980s onwards, biodiversity-rich developing countries had been pushing for greater protection of their interests in patented products that were developed by multinational companies but based on freely available genetic resources from the South. Developing countries were therefore trying to commit developed countries to ensuring that multinationals would share the benefits of their use of genetic resources, particularly through technology transfer and a change in intellectual property rights (Porter, 1992).

Given the centrality of patent protection to modern biotechnology, the US biotechnology and pharmaceutical industries opposed proposals for benefit sharing in the CBD, pointing to their threat to existing intellectual property rights (Hoyle, 1992). Their concerns fell on fertile ground in the Bush Administration. Senior White House officials held reservations about the legal implications of a binding international biodiversity agreement and feared that it would undermine their efforts to roll back anti-business regulations (Hopgood, 1998: 172). US negotiators were concerned that the treaty would entail an open-ended financial commitment, that its benefit-sharing provisions would violate existing intellectual property rules, and that the biotechnology industry would be burdened with unnecessary regulations (Haas et al., 1992: 14; Porter, 1992). The US delegation succeeded in extracting several concessions from the developing countries, including on biotechnology regulation and intellectual property protection, and although other industrialized countries considered these sufficient to accede to the convention, the US stood firm in its opposition. When the final draft text was tabled in Nairobi in May 1992, over 150 countries, but not the US, signed the accord (Porter, 1992). This was to weaken the US's longer-term position.

The outcome of the CBD negotiations left the US isolated on the diplomatic stage, but industry representatives and White House officials felt it was worth paying this price. Through the Industrial Biotechnology Association, which represented some 80 per cent of the US biotechnology market, the US biotech sector had vigorously objected to the convention during the talks and expressed support for the Administration's decision not so sign it, even though it had little time to study the final compromise text (Porter, 1992). This was in contrast to the position of biotechnology firms outside the US. While they were less engaged in the industry lobbying efforts during the talks, European and Japanese industry associations did not share the same negative assessment of the convention and were less concerned about the legal implications of the convention for intellectual property rights protection (International Environment Reporter, 1992f).

By refusing to sign the CBD, the US may have avoided exposing itself to legal commitments that ran counter to its preferences, but it soon became apparent that this also weakened US influence in the further development of the global biodiversity regime. For a compromise had been found at UNCED in the form of a commitment in Article 19 of the CBD to consider the need for and modalities of a separate protocol on biosafety. The US was now relegated to the status of an observer as soon as the Conference of the Parties (COP) set upon deciding on a biosafety mandate. Although the US could participate in the COP meetings just like other non-parties, it was barred from voting and often found itself excluded from the more politically charged discussions. This was to render the US less effective in the future biosafety talks, and also limited the influence of the biotechnology industry whose main supporter the US had been.

The diplomatic isolation that the US suffered at UNCED led to a strategic rethink once the Clinton Administration entered office in 1993. The US biotechnology industry also revised its strategy amidst fears that future rule-making within the CBD framework could harm its interests. With the encouragement of Vice President Al Gore, three US firms (Merck, Shaman Pharmaceuticals and Genentech) and three NGOs (World Resources Institute, World Wildlife Fund, and Energy Studies Institute) formed a private initiative to study the obstacles to acceding to the CBD. Based on a more sober assessment of the legal issues that the CBD presented, the group concluded that the US could sign the convention without infringing intellectual property rights, as long as it was able to add an interpretative statement to safeguard corporate interests. The three corporations managed to win the support of the Pharmaceutical Manufacturers Association for the new Administration's stance on the CBD, but stated in their draft interpretative statement to the president that they saw no need for a protocol on the international trade in GMOs. In the end, it was not opposition in US boardrooms but in the US Senate that prevented the country from joining the biodiversity regime. Despite winning the support of leading industrialists, Clinton failed to secure ratification of the treaty on the Senate floor, where Republican opposition ran high against what Senators saw as a signing away of sovereign rights and a violation of intellectual property rights (Munson, 1995).

Agreeing on a biosafety negotiation mandate

The CBD entered into force in December 1993. A year later, the first Conference of the Parties (COP-1) was convened in Nassau, the Bahamas, at which developing countries put the question of a biosafety treaty back on the agenda. The leading industrialized countries remained as sceptical on this question as they were at UNCED two years earlier. Biotechnology industry groups on both sides of the Atlantic opposed international regulations, but few paid any serious attention to the CBD debates. The EU was unable to arrive at a common position. Europe's main biotechnology countries remained unconvinced of the need for binding international rules and recommended instead a set of voluntary safety guidelines that were being drafted at the initiative of Britain and the Netherlands, which became the UNEP International Technical Guidelines for Safety in Biotechnology (UNEP Guidelines). COP-1 concluded without any progress on the biosafety question, having focused mainly on institutional matters in the biodiversity area.

The failure to endorse a mandate for biosafety negotiations meant that the EU's international position was increasingly out of sync with its domestic approach to biosafety. With the 1990s GMOs directives the EU had created its own set of precautionary safety regulations, but continued to question the need for similar rules at the international level. For a short period in the early 1990s, it seemed as if anti-regulatory forces were gaining in influence again in Brussels (Shackley and Hodgson, 1991; Torgersen et al., 2002: 50–60). Indeed, shortly after 1990, pressure grew on the EU to provide a more fertile environment for biotechnology, and various efforts got under way to review and harmonize existing GMO regulations (Hodgson, 1991). Europe's new regulatory framework was one of industry's main sources of frustration but also provided it with a focal point for political mobilization and organization. Unlike the various anti-GMO campaign groups that were mostly nationally organized, industry was quick to put its lobbying effort on a European footing and saw the power balance temporarily shift in its favour (Grabner et al., 2001: 17). After the creation in 1989 of SAGB, the various national biotechnology associations joined forces and in 1991 formed the European Secretariat of National Bioindustry Associations (Cantley, 1995: 635), and in 1996 created EuropaBio, an umbrella body that has since developed into the main European biotechnology group. Their demands for a less restrictive regulatory environment were boosted by the growing perception that EU efforts to close the global technology gap were faltering. Despite the fact that Europe had increased spending on research and development (R&D), the US industry remained the most dynamic worldwide. By the mid-1990s, there were twice as many start-up companies in the US than in Europe, US companies were spending ten times as much on R&D than their European competitors, and their revenues were nine times higher than those in Europe (Paugh and Lafrance, 1997: 101).

But the industry-friendly turn in Brussels was short-lived, as international developments took on a dynamic of their own and fed back into European policy-making. While European industry lobbying concentrated on the reform of EU legislation, the EU came under increasing international pressure to agree to the start of biosafety negotiations. The issue of whether to create a biosafety protocol was again on the agenda of the Second Conference of the Parties (COP-2), in Jakarta in 1995. Despite the setback at the first COP meeting, developing countries reopened the debate on biosafety at COP-2 (La Vina, 2002). The US delegation continued to oppose biosafety negotiations, as did the US biotechnology industry. Positions within the EU had moved little since UNCED, but a compromise position was beginning to emerge. The EU Council had adopted a negotiation mandate for COP-2 that balanced the diverging views of member states, envisaging a two-track approach of supporting voluntary biosafety guidelines while agreeing in principle to talks on a biosafety treaty (Bail, Decaestecker and Jørgensen, 2002: 169). European Commission representatives advocated this as the only consistent and legitimate position to take for the EU, given its commitment to precautionary regulation in Europe. As a large number of parties to the CBD were now willing to open negotiations, the EU consented with the majority view without attaching too much importance or urgency to the process. COP-2 adopted what came to be known as the Jakarta Mandate for a biosafety treaty, and established an Open-Ended Ad Hoc Working Group on Biosafety to discuss the regulatory options and prepare a draft decision for adoption at a future COP meeting.

Business played only a marginal role in this process of agenda-setting. Not many companies took notice of the discussions in CBD meetings, and those that did attached no greater importance to the adoption of a mandate to negotiate a biosafety treaty. At this point, it was still unclear what the content of this treaty might be, or whether it would ever come into existence. As is common in such situations, industry groups took a reactive approach and preferred to wait and see what would come out of this process. The biotechnology industry was also handicapped by a weak international presence, which in itself reflected the diversity and in many cases small size of biotechnology firms. The comparatively more advanced biotechnology sector of the United States did express opposition to CBD plans for a biosafety treaty, but its voice in the international context was constrained by the fact that the US could not participate in the talks as a CBD party. European biotech firms were mostly focused on the European regulatory debate, with the EU's 1990 regulations their main concern, and did not therefore consider international biosafety developments a priority. In this context, a loose coalition of developing country representatives, environmental campaigners and regulators from the North were able to create an international biosafety agenda and a mandate to start elaborating the details of a future biosafety regime.

Negotiating the Cartagena Protocol on Biosafety

As the international biosafety negotiations gathered pace in the second half of the 1990s, business actors involved themselves more closely in the process. The firms that followed the international process most actively from its start were those of the biotech sector. Their products, genetically modified seeds, were one of the key targets of proposed biosafety regulations. Over time, as the implications for commodities trade became clearer, the farm trade sector also came to develop a lobbying strategy and a stronger presence in the talks. Most firms were opposed to stringent biosafety rules and preferred non-binding safety standards. But their cause was weakened by the uneven corporate involvement in the talks, and as the political momentum for an international regime grew, cracks in the business front served to weaken the anti-regulatory business lobby.

At least two of the common types of business division could be observed in the biosafety negotiations. First, North American and European biotech companies differed in their lobbying styles and tactics, with the former taking a more assertive anti-regulatory stance. Given the high degree of transatlantic industry integration, however, these differences never threatened to undermine the underlying unity in outlook. Second, a more important cleavage arose between biotechnology firms in agriculture, the main target of the protocol, and those operating in the pharmaceutical sector, whose interests were only marginally affected. With both these biotechnological sectors going their separate ways following the industrial restructuring of the late 1990s, the emerging divisions between them came to severely undermine the former's lobbying clout in the biosafety talks.

Further down the GM food production network, seed distributors, farmers, agricultural traders, food producers and food retailers could also be affected by the proposed biosafety rules, although the further removed companies were from the biotech seed sector, the more difficult it was for them to identify clear corporate interests in the biosafety area. Out of this diverse business group, agricultural traders' interests were most directly affected, and as soon as the trade dimensions of the biosafety protocol became clearer half way through the talks, they also developed a strong lobbying effort. The arrival of agricultural interests groups on the diplomatic scene helped to weaken the widespread perception that business opposition to the biosafety regime was narrowly driven by the interests of a handful of multinational biotech firms. At its height, the newly created Global Industry Coalition (GIC) came to represent 2,200 companies in 130 countries (PR Newswire, 1999), including biotechnology and farming and food producer interests. Maintaining unity among such a wide range of business interests proved difficult at times, however, and in the end it was agricultural trading firms that remained most steadfastly opposed to the adoption of the biosafety protocol, long after the biotech sector had come to accept the need for the agreement.

A slow start in the Biosafety Working Group

The international biosafety talks kicked off with the first meeting of the Biosafety Working Group (BSWG) in Aarhus, Denmark, in July 1996 (BSWG-1). The meeting attracted little media attention and was attended by only a small number of industry groups. Although a formal negotiation mandate had been issued by COP-2, the scope and nature of such an agreement – should it ever be adopted – were still fuzzy. It was therefore unclear how such a treaty might affect specific business interests, and given the lack of support for a biosafety instrument among the leading biotechnology countries, it was entirely conceivable that these talks might not lead anywhere. It would, of course, not have been the first time that efforts to create binding international environmental rules might falter. Only four years earlier, in 1992, plans for an international treaty to protect forests had been watered down to a mere declaration of non-binding Forest Principles (Davenport, 2005; Humphreys, 2001). Given the high degree of uncertainty surrounding the biosafety talks, therefore, most companies in the global food production network did not feel the need to be too closely engaged in the process.

If anything, the early BSWG process confirmed business leaders in their belief that they could afford to sit back and wait. The few industry observers present at the BSWG-1 went away feeling disillusioned about the diplomatic process (Reifschneider, 2002: 274). The first round of talks produced little more than a shopping list of regulatory elements that delegates wanted to see included in a future agreement. Under the leadership of Ethiopia's Tewolde Egziabher, developing countries put forward far-reaching proposals that sought to establish a comprehensive regulatory system covering all types of GMOs, which industry observers were quick to dismiss as unwarranted and unworkable. But the business case was not helped by the fact that only three business sectors (seeds, forestry and aquaculture) had sent observers to the talks. High turnover among their ranks meant that the few industry representatives who were present developed no rapport with the governmental delegations. According to Laura Reifschneider, the only industry lobbyist to be present throughout the entire negotiation process, the business sector suffered from 'imperfect coordination' (2002: 274) in the early phase.

BSWG-1 did not signify the beginning of the negotiation process as such; it was only the first in a series of meetings at which delegates put regulatory options on the table without seeking to resolve their differences (see Falkner, 2002 on the entire process). This was to some extent a reflection of the novelty and complexity of the proposed international treaty. Many delegations had not yet formed a clear view of what their preferences were and were therefore simply creating a 'wish list' of regulatory measures from which they hoped to derive a consensual draft treaty (International Environment Reporter, 1996h). It was also the strategy chosen by the chairman of the BSWG process, Veit Koester of Denmark, who wanted to promote the search for a broad consensus on the major treaty components and repeatedly reminded delegates that they were still in the pre-negotiation stage (Koester, 2002). This open-ended, at times seemingly unstructured, nature of the international process made it difficult for industry lobbyists to engage with it and explains some of the frustration they experienced during the talks.

Although the basis for the BSWG discussions had been laid in the 1995 UNEP Guidelines and a list of regulatory options that had been drafted at a meeting of biosafety experts in Madrid in July 1995, delegates arrived at the first BSWG meetings with a vast array of different regulatory proposals. Most of the divisions that characterized the negotiations on the CBD and the Jakarta Mandate resurfaced in subsequent BSWG meetings. Broadly speaking, a large group of developing countries argued for a comprehensive regulatory regime that would cover all types of GMOs and products resulting from genetic engineering; give importer nations extensive powers to assess and potentially block GMO imports, including on socio-economic grounds; and include a liability regime and financial mechanisms to support capacity-building initiatives. In contrast, most industrialized countries, which had their own national regulations in place, favoured a more limited approach focused on transboundary movements of GMOs and opposed references to socio-economic considerations and the creation of a liability regime. While there was therefore a broad North–South division in the talks, significant differences also existed within these large groups. Some developing countries, such as Argentina and to some extent Brazil, placed greater emphasis on making biosafety rules consistent with international trade rules. And among the industrialized countries, the US delegation continued to question the very need for binding international rules and continuously argued for a minimal, non-binding and World Trade Organization (WTO)-consistent solution. Within Europe, the Nordic countries showed sympathy for the developing countries' position, but given scepticism by Germany and others, the EU was still in search of a common position (Bail, Decaestecker and Jørgensen, 2002; Egziabher, 2002; Enright, 2002).

The second and third BSWG meetings, held in Montreal in May and October 1997, continued in the same mode with further discussions on the proposed treaty elements, many of which remained contested. A basic treaty structure was beginning to emerge by the time of BSWG-3, with two sub-working groups and two contact groups now focusing on clustered sets of issues, but it also became clear just how contentious certain issues were (Falkner, 2002: 10–13). Industry groups had grown increasingly frustrated at what they saw as a near total neglect of the economic and technical realities of GMO and agricultural trade. Due to the US delegation's limited role as an observer group, many North American companies looked to Canada for support in presenting their case inside the diplomatic proceedings. Although Canada has been a strong supporter of the CBD – it was the first industrialized country to ratify the agreement, and Montreal was chosen as the home of the convention's secretariat – the country came to align itself with the US position as Canadian farmers began to adopt GM crops after 1995. The Canadian delegation later even came to chair the Miami Group, the pro-trade alliance of agricultural exporter countries that opposed many of the Cartagena Protocol's key provisions (on the Canadian position, see Ballhorn, 2002).

Up until 1998, biotech firms dominated the corporate lobbying efforts in the biosafety talks. Did transatlantic corporate differences shape their political strategies, as has been the case in other environmental negotiations (see chapters 3 and 4; Levy and Newell, 2000)? Participants in the Cartagena Protocol talks have attested to some regional variation in the role played by biotech firms. The political strategies and lobbying styles of US and European biotech companies reflected the national context in which they usually operated. While US corporate representatives adopted a more confrontational and robust approach, challenging the demands for a process-based and precautionary system of risk regulation, European lobbyists tended to strike a more conciliatory tone, emphasizing the need to cooperate with regulators and to mould biosafety rules with business needs in mind (International Environment Reporter, 1996i). This impression was shared by environmental NGOs who participated in the biosafety negotiations. Richard Tapper, an advisor to the Worldwide Fund for Nature during the biosafety talks, remarked that European biotechnology representatives 'were more open to constructive dialogue than some of their US counterparts' (2002: 271).

But did these regional differences in business lobbying translate into a significant business conflict that would affect business strategies in biosafety politics? There is little evidence to suggest that the biotech industry was split along regional lines or that the different lobbying tactics they chose in the talks were rooted in more fundamental differences in political outlook or strategy (Clapp, 2007). The main reason for this is the high degree of internationalization in biotechnology that has helped to blur regional differences, both in terms of global commercial strategies and industry structures. For one, biotech firms in North America and Europe have been developing products for the same markets around the world. They have developed similar biotech products and have sought regulatory approvals for their crops in the same markets. They thus had to face the same political constraints in the markets in which they sought to introduce GM crops. They had to adjust their strategies to local and regional conditions, but maintained the same overall strategic outlook. As Levy and Newell argue, 'despite procedural and institutional influences on the way businesses have pursued their interests and different degrees of exposure to social and political concerns about GM foods, biotech companies in both Europe and the United States have maintained similar positions' (2000: 14).

With regard to industry structure, the growing internationalization of biotech research and development has also rendered regional differences less relevant. The series of mergers and acquisitions described at the beginning of this chapter has led to a dramatic concentration of agri-biotechnology in the hands of only five European and North American firms. Even though the remaining firms retain strong links to their domestic markets in the US, Germany and Switzerland, their R&D, field testing and commercialization efforts are highly internationalized. European biotech firms such as Syngenta and Bayer have a strong presence in the North American market and are in close competition with Monsanto in overseas markets in Asia, Africa and Latin America. Because of industrial internationalization, it makes only limited sense to speak of Bayer and Syngenta as representing a European biotechnology interest. Or, as one US industry lobbyist put it, 'there was no European industry position as such' in the negotiations on the Cartagena Protocol (interview with Karen Anderson, Biotechnology Industry Organization, 16 July 2001).

Agricultural commodities move centre stage in the biosafety talks

The start of the biosafety negotiations in 1996 coincided with the first large-scale commercial planting of GM crops. That year, GM crops were grown only on an estimated 1.7 million hectares, but the adoption rate of the new technology shot up quickly in subsequent years, with the global GM crop area rising to 11 million hectares in 1997, 27.8 million in 1998, and 39.9 million in 1999 (James, 1999). Initially, GM crop production was heavily concentrated in only three countries, the US, Canada and Argentina. Given the ubiquity of the first generation of GM crops (corn, soybean and cotton) in international trade, however, many other countries soon had to grapple with the biosafety issues that GM crop and food imports raised. Thus, by 1998 GM content became more widespread in international farm trade, and the negotiators focused more explicitly on the commercially sensitive question of how to deal with agricultural commodities trade in the biosafety protocol.

By this time, agricultural traders had also woken up to the potential threat that the emerging biosafety regime posed to their business interest. North American and international farm trade associations started to be more involved in the biosafety talks and sought to highlight the implications that biosafety import rules would have on agricultural trade practice, particularly if stringent identification requirements would be adopted. Many felt that the BSWG delegates, most of which were scientists or regulators from environmental and health ministries, did not adequately understand the reality of global commodities trade. Farming and trading associations sought to enlist the support of agriculture and trade ministries at the domestic level and lobbied delegates internationally. In the end, most developing countries firmly stuck to the negotiation strategy of the Like-Minded Group, and even though some of the larger agricultural trading countries such as Brazil started to waver towards the end phase of the negotiations, the developing country alliance managed to avoid defections until after the conclusion of the talks (Nogueira, 2002; Falkner and Gupta, 2004: 11).

By the time of the fourth and fifth BSWG meetings in February and August 1998, the trade dimensions had moved centre stage in the biosafety talks (International Environment Reporter, 1998d). In July of the same year, the United States initiated the formation of the so-called Miami Group (including Argentina, Australia, Canada, Chile and Uruguay), which took a decidedly more pro-trade position and became the business community's closest ally in the talks. This move soon led to a wider realignment of the negotiation groups, as the regional groupings usually favoured in UN settings no longer served as an appropriate tool for aggregating country interests. This process was already under way at the sixth and final BSWG meeting in Cartagena, Colombia, in February 1999, as well as the extraordinary meeting of the Conference of the Parties (ExCOP) that followed immediately from BSWG-6. Once it became clear that three members of the G77 (Argentina, Chile and Uruguay) had joined the Miami Group, the developing country bloc regrouped itself as the pro-regulatory Like-Minded Group in 1999. At the instigation of the ExCOP chair Juan Mayr, other countries also formed negotiating groups that represented more homogeneous sets of interests: the European Union, which had acted as a single negotiating group from the beginning and was now developing a more coherent and proactive role in the talks; the Central and Eastern Europe Group, which comprised Russia and East European countries, and continued to play a mostly marginal role; and a new grouping consisting of all those countries occupying the middle ground in the talks, which became known as the Compromise Group (Japan, Korea, Mexico, Norway and Switzerland, later joined by Singapore and New Zealand) (Falkner, 2002: 17).

On many of the critical issues in the talks, the Like-Minded Group and the Miami Group occupied the opposite ends of the political spectrum. The former demanded that all GMO types, including products derived from GMOs, were covered by the regulatory instrument of the advance informed agreement (AIA), which meant that exporters of GMO seeds and commodities needed to seek the prior approval of the importing country. The Miami Group, in line with the Global Industry Coalition, rejected this blanket application of the AIA procedure and particularly opposed the inclusion of agricultural commodities. The developing countries sought to expand the grounds on which importers could reject GMO shipments, including socio-economic considerations and precautionary reasons. However, the Miami-Group would only accept a narrow, science-based, definition of risks that should inform importer decisions, and insisted that this be consistent with existing international trade obligations. The former wanted to create a liability regime that forced GMO producers and exporters to compensate importing nations should GMO releases cause damage to the environment. The latter, supported by the EU on this issue, rejected this as unworkable and unnecessary (International Environment Reporter, 1998a).

The fourth and fifth BSWG meetings did little to resolve the underlying conflict. As Val Giddings, vice president for food and agriculture for the Biotechnology Industry Organization, commented in 1998, 'the trenches are deeper and the ramparts higher' (International Environment Reporter, 1998d). US industry groups firmly supported the US stance and praised the delegation for remaining firm despite having been isolated in the early talks. Daniel Amstutz, head of the North American Export Grain Association, urged the US government to seek exemption from the protocol of agricultural commodities that are destined for processing or consumption, arguing that segregation between GM and non-GM varieties was 'impossible' (International Environment Reporter, 1999a: 61). At the same time, however, he conceded that his sector could live with a requirement to identify shipments that 'may contain' GMOs, in response to European demands for comprehensive and specific identification of GMO content in all international trade (ibid.).

The final BSWG meeting in February 1999, which was to prepare a compromise text for adoption at the subsequent ExCOP, made little progress in narrowing the remaining differences. When BSWG chair Koester presented his own compromise proposal in the form of a chair's text, no consensus could be found on this and the text was submitted in revised form to the ExCOP meeting where negotiations continued in an equally confrontational manner. As before, seemingly unbridgeable differences existed between the Like-Minded Group and the Miami Group. What had changed, however, over the last few BSWG meetings was that the European Union had come out in support of a strong, precautionary, biosafety treaty and provided an important leadership role in the final rounds of negotiations. Towards the end of the ExCOP meeting, the EU put forward a compromise package aimed at reaching an agreement, which entailed concessions to the Miami Group as it offered to postpone a decision on the question of how to treat trade in commodities; and to the Like-Minded Group in the form of a provision that would allow the future inclusion of commodities in the AIA procedure. Recognizing that this was the best hope for reaching a compromise with the Miami Group, the developing countries reluctantly came out in support of the EU position. But although the other groups also approved the EU package, the Miami Group felt unable to agree to the new terms. The ExCOP meeting was suspended without a result, and the parties agreed to reconvene in 2000 to seek a new compromise deal (International Environment Reporter, 1999b; Falkner, 2002: 15–18).

Domestic–international linkages in the EU's changing biosafety policy

The change in the EU's position from disunity and indifference in the mid-1990s to partial leadership in 1999 represents one of the key turning points in the history of the biosafety negotiations. Without it, the successful conclusion of the talks would have been highly unlikely, and although the Like-Minded Group was the key *demandeur* for a strong international biosafety instrument, it was the EU's eventual leadership role that played a critical part in bringing about a compromise in January 2000. Given the EU's prominent position as a host to leading biotechnology firms and as a major market for agricultural products, it had the necessary economic and political clout to extract a compromise deal from the Miami Group. What brought about this change in the EU's international position? Only a closer look at the domestic political economy of biotechnology in Europe can reveal the origins of this policy shift (on the following discussion, see Falkner, 2007b).

The EU's early involvement in international biosafety politics was characterized by a lack of clear interest and unity. Until around 1998, when the trade dimensions of the protocol became more pronounced, Europe was split between those countries, such as Austria, Denmark, Norway and Sweden, that were more sympathetic towards developing countries' demands, and others such as Britain and France that favoured the non-binding UNEP Guidelines as a model for the protocol. Germany under the conservative government of Helmut Kohl was also opposed to a binding biosafety protocol. The two-track approach adopted by the EU in 1995, which involved the parallel development of the voluntary safety guidelines and the elaboration of a biosafety treaty, merely sought to bridge the diverging views of key EU states but was unable to resolve their underlying disagreements.

Between 1997 and 1998, however, a fundamental change was under way in Europe that would transform the EU's role in the negotiations, align it more closely with the position of the Like-Minded Group and give rise to European leadership in the critical end phase of the negotiations. Two developments were behind this transformation: the arrival of GM crops in international trade, which helped to mobilize a European-wide anti-GMO campaign and increased the political salience of the biosafety negotiations; and a corresponding shift in the EU's domestic GMO policy towards greater precaution that helped to create a more united, and proactive, international position.

In the autumn of 1996, Europe received the first shipment of GM crops from the US, which attracted widespread media coverage, led to direct actions and boycotts by environmental campaign groups and fuelled fears about food safety among European consumers (International Environment Reporter, 1996j; Pollack and Shaffer, 2005: 21). Several authors have explained the strength of Europe's anti-GMO movement with reference to a wider crisis in European food safety (Pollack and Shaffer, 2005; Vogel, 2003), particularly in the wake of the BSE crisis. Although BSE, or 'mad cow disease', was unrelated to plant genetic engineering, it cast a shadow over the new technology. Activist groups such as Greenpeace and Friends of the Earth orchestrated protests against experimental GMO planting and GM food sales around Europe. Unlike in the United States, where the public remained largely indifferent to GM food, European opinion polls revealed rapidly rising anti-GMO sentiment among a public concerned more with food safety than agricultural productivity (Torgersen et al., 2002: 61–74).

The GMO controversy created a dilemma for EU authorities. Whereas the Commission believed its regulatory system was working well, some EU member states' governments (e.g. Austria, France and Luxembourg) reacted nervously to rising anti-GMO sentiment and demanded a temporary halt to new GMO authorizations (International Environment Reporter, 1998c). Based on Directive 90/220, the Commission was able to approve a new GM maize variety in January 1997 despite failing to win the support of member states. But the GMO authorization process eventually collapsed in 1999 as more and more governments invoked the Directive's safeguard clause, thus preventing the implementation of GMO approvals in member states (International Environment Reporter, 1998b, 1998e; Pollack and Shaffer, 2005: 23–4). European biotechnology policy entered a 'perfect storm' in 1998/99 as it came under simultaneous attack from different sides: parallel anti-GMO campaigns by environmental and consumer groups sprang up in several member states; the European Parliament called for a strengthening of regulations; and food retailers withdrew already authorized GM food from the market. All the while, the European biotech sector struggled to overcome its fragmented industry structure and failed to win over European farmers to the new technology (Bernauer, 2003: 80–6). The EU had no choice but to seek to calm the situation by introducing a *de facto* moratorium on GMO approvals while it sought to revise its regulatory framework (Skogstad, 2003: 327–30). By 1999, at a crucial point in the international negotiations, precaution had won over trade liberalization and competitiveness concerns in shaping Europe's political identity in biotechnology regulation. Lacking allies in the farming and food production sector, European biotech firms alone were unable to prevent a further deterioration of the EU's regulatory situation.

The domestic shift in Europe had two important consequences for international biosafety politics. First, the EU's increasingly restrictive stance on GMOs sent an important signal to those developing countries still deciding whether or not to adopt GM technology, reinforcing existing biosafety concerns there. Moreover, with the European food market temporarily closed to GM crop imports, developing country exporters in Africa and Asia feared that their own farm products would be barred from the European market if they introduced GMOs domestically. The European GM debate was therefore closely watched in the developing world, and Europe's GMO moratorium further hardened anti-GMO positions especially in Africa and Asia (Clapp, 2006; Falkner, 2006; Gupta and Falkner, 2006).

Second, the EU's negotiation role changed as the international context gained political salience in Europe and the negotiations became a test case for the EU's ability to withstand North American pressure. American threats to open Europe's gates to GM products by launching a legal challenge at the WTO only served to underline the importance of a strong international treaty that would lend legitimacy, if not full legal cover, to Europe's regulatory framework. From 1998 onwards, EU negotiators pushed for a more precautionary approach in the biosafety protocol, helped also by the fact that Germany abandoned its more sceptical stance on the protocol after the social democratic and green parties won the 1998 national election (International Environment Reporter, 1998a, 1998f). The fact that the focus of the talks had shifted to the regulation of agricultural commodity shipments had promoted a more united EU negotiation role, as trade issues were traditionally negotiated by the Commission on behalf of the Union. In this case, however, the Commission's Environment Directorate-General, not the Trade Directorate-General, led the negotiations and pushed for an international agreement that closely mirrored its own EU-wide GMO regulations.

The EU's key demands in the final round of negotiations, from 1999 to 2000, concerned three elements of the treaty that were fiercely resisted by the US-led group of agricultural export countries (Falkner, 2002):[2]

- The precautionary principle, allowing importing countries to restrict GMO trade under conditions of scientific uncertainty. This was seen as a major plank in the defence against potential WTO legal challenges.
- Clear rules on identification of GM content in international agricultural trade, to support domestic regulatory authorities.

- The mutual supportiveness of the biosafety treaty and international trade rules, in order to prevent agricultural exporters from challenging GMO import restrictions at the WTO.

While European negotiators aligned themselves with the developing countries to reach an agreement on these and other elements of the treaty, they did not go along with all their demands. On the whole, EU negotiators failed to offer leadership in the negotiations on issues that went beyond what was contained in the EU's domestic regulatory framework. The Like-Minded Group of developing countries had argued from the beginning for a more comprehensive scope of the protocol than both the EU and the United States were willing to accept, to cover not only GM seeds and commodities but also products derived from GMOs. Demands for a liability regime, which would oblige GMO exporters to compensate importing nations for any future damage inflicted from the release of GMOs, were also met with resistance by the US and scepticism by the EU (International Environment Reporter, 1998a). Finally, the inclusion of GMOs that were pharmaceuticals for humans was blocked by both the US and the EU. On the latter, the EU argued that pharma-ceutical GMOs fell outside the remit of the protocol as they were covered by other international agreements and were not designed to be released into the environment, although recent advances in the medical use of GM crops and food products have started to blur this distinction. This, therefore, was not simply a question of the technical delineation of one regime context from another, but a question of political expediency. Pharmaceutical companies had been lobbying hard at the biosafety negotiations to have their products exempted from the protocol's scope and found receptive ears among the EU and US negotiation teams.

Emerging business conflict in the biotechnology sector

The failure to adopt a biosafety protocol at the ExCOP meeting in Cartagena, February 1999, had two immediate effects. First, it helped to dramatize the underlying North–South and US–EU conflicts on GMO regulation and created greater public interest in the biosafety negotiations. Second, it strengthened the resolve of all parties to try to overcome the remaining differences and reach a compromise in a resumed ExCOP meeting that was scheduled to take place before March 2000. The renewed interest in a successful conclusion of the talks was helped by parallel events in other global policy areas. The year 1999 saw a dramatic increase in campaigns against GM food in Europe and elsewhere, with environmental NGOs uprooting GM plants and consumer groups boycotting GM food on supermarket shelves (Weiss, R., 1999). Opposition to biotechnology also featured prominently in the anti-globalization movement that made front page news with violent protests against international bodies such as the G8 and the WTO. The widely reported collapse of the December 1999 Ministerial Meeting of the WTO in Seattle, which saw street fighting between anti-globalization campaigners and the police outside the main conference buildings, signified to many the wider legitimacy crisis that international policy-making had fallen into. In this climate of a deepening crisis of international governance, many negotiators in the biosafety area felt compelled to renew their efforts to reach an agreement, not least as a demonstration of the capacity of international diplomacy to respond to popular concerns (Gupta, 2000).

Policy-makers were not the only ones who felt the pressure of growing popular discontent. Mounting civil society opposition to genetic engineering reached a peak in 1999 and began to dent, at least for the moment, the commercial growth prospects of biotechnology. With a *de facto* moratorium on GM crop authorizations and imports in place and supermarkets removing GM food products from their shelves, the European market was temporarily closed to new biotechnological developments (see the following section, below). Many European biotechnology firms lost investor confidence in a climate of public hostility and wound up their agricultural research programmes (Hodgson, 2000; Nature Biotechnology, 1999). While Europe remained the centre of anti-GMO activity, protests increasingly spread to other parts of the world, including the US (Fox, 1999). In the developing world, concern over GM crops focused on the development of seeds that produce infertile crops, thus preventing farmers from propagating the crops themselves and forcing them to buy seeds for every growing season. Amidst mounting criticism of this so-called 'terminator technology', Monsanto announced in October 1999 that it would no longer seek to commercialize the technology. The announcement was widely interpreted as a sign that the company needed to be seen to be listening to its critics (Niiler, 1999a, 1999b).

Pressure on biotechnology firms was also building as investors and financial analysts became increasingly wary of the bad publicity the GM food controversy caused for the entire biotechnology sector. In May 1999, Deutsche Bank analysts released a widely noted report titled 'GMOs Are Dead', in which they warned investors of the political risks surrounding genetic engineering in agriculture and advised them to sell shares in leading agri-biotechnology firms (Brown and Vidal, 1999). Although these predictions have not come true and shares in firms such as Monsanto have since recovered, the report sent shockwaves through the institutional investment community and reinforced the trend towards separating pharmaceutical and agricultural biotechnology businesses.

The effect of this restructuring trend was felt most vividly in late 1999 and 2000, which saw some of the biggest mergers in the biotech sector. In Europe, two of the leading firms in the biotechnology sector, Novartis and AstraZeneca, decided in 1999 to spin off their agricultural businesses, which accounted for 26 per cent and 16 per cent respectively of their total sales in the first nine months of that year. The newly formed company Syngenta was to combine the agri-biotech spin-offs, leaving Novartis' and AstraZeneca's core pharmaceutical business untainted by association with GM foods (Niiler, 2000a). Shortly thereafter in 2000, Monsanto, which had combined agricultural and medical biotechnology but became the main target of anti-GMO food campaigns (Niiler, 1999b), merged with the pharmaceutical firm Pharmacia & Upjohn. Again, Monsanto's agricultural biotechnology business, which accounted for nearly 48 per cent of the company's total sales in 1999, was separated out from its pharmaceutical business and restructured to operate independently under Monsanto's name (Martinson, 1999; Niiler, 2000b).

The reorganization of the biotechnology industry led to a growing divergence in the political strategies of biotechnology firms. Differences in commercial and political interests had already been noticeable throughout the 1990s. In fact, the diversity of biotechnology firms, spanning pharmaceuticals, agriculture, environmental remediation and industrial processes, had been holding back lobbying efforts, as sector-wide organizations struggled to bridge the differences between their commercial and strategic orientations (Dorey, 1999). Now that the pharmaceutical and agricultural sides of biotechnology were increasingly going their own ways, plant and animal biotech firms found their lobbying clout significantly reduced. This was particularly noticeable in Europe, where the restrictive environment for GM food development had hindered expansion of the agri-biotech sector. A European Commission Joint Research Centre study of 2003 found that as a result of the EU moratorium on GMO authorizations and hostile consumer reactions, the number of plant field trial applications in the EU had fallen by 87 per cent between 1998 and 2002, and that more than half of small and medium-sized and two-thirds of large biotech companies cancelled research projects over this period (Lhereux et al., 2003). The growth in biotechnology that Europe experienced during this time happened only in the pharmaceutical sector. Agriculturally oriented biotech firms had become marginal players in the industry and could no longer count on support from the pharmaceutical sector. By the late 1990s, the latter were seeking to insulate themselves from the negative fallout of the GM food debate, rather than tackling GMO critics head on as part of an integrated political strategy of all biotech firms.

The growing divergence between agricultural and pharmaceutical biotechnology had an indirect, though important, effect on the final stage of the biosafety negotiations. With the agricultural biotechnology sector increasingly embattled and under pressure from activists, consumers and investors, its claim to represent a major future growth sector looked more questionable than ever before. Social protest and business conflict had served to reduce the agri-biotech sector's structural power and damaged its legitimacy in the regulatory discourse. And while the biotechnology sector never lost the support of the US government, its position in European policy networks and relational power vis-à-vis EU negotiators was severely curtailed.

The impact of business conflict within the biotechnology sector became evident in the final round of the biosafety negotiations, after the collapse of the 1999 Cartagena meeting. Throughout the talks, pharmaceutical companies had taken part in the overall biotechnology industry lobbying ever since the Global Industry Coalition was created in 1998. As the talks progressed, they increasingly focused their lobbying effort on the question of whether pharmaceutical GMOs would be included in the treaty. Having secured EU support at BSWG-5 for the proposal to exclude such GMOs by listing them in an annex of exemptions to the protocol, they succeeded in distancing the pharmaceutical sector from the regulatory dispute over GMO trade (Marquard, 2002: 294). This solution was discarded, however, by the Like-Minded Group shortly after the collapse of the Cartagena conference in 1999. The developing countries reopened the question of including pharmaceuticals in the protocol and demanded a renegotiation of this provision, arguably to bolster their bargaining position in the concluding round. Having seemingly secured an exemption from the treaty, the pharmaceutical industry was yet again faced with the threat of pharmaceutical GMOs being subjected to the treaty's advance informed agreement procedure. Helen Marquard, who headed the UK delegation, describes the uproar that followed:

> This caused a noticeable stir in the pharmaceutical industry in Europe and North America. It had supported the protocol, seeing it as furthering technology transfer, but it wanted to ensure that it would not hinder the movement of necessary medicines. An AIA procedure would, in its view, have done so. (Marquard, 2002: 295)

Rather than closing ranks with agri-biotechnology firms in opposing the treaty altogether, leading pharmaceutical firms focused their last-minute lobbying in the final negotiation round in Montreal on the single issue of excluding pharmaceutical GMOs from the scope of the treaty. Neither US nor EU negotiators had an interest in covering both biotech sectors under this treaty, and the pharmaceutical industry was given assurances that the leading industrialized countries would not budge on this issue. Having thus secured their desired exemption clause, which was enshrined in the treaty in Article 5, some pharmaceutical industry lobbyists were able to leave the Montreal conference even before it reached its conclusion. The agricultural biotechnology sector was left to bat alone.

In this way, the disintegration of agricultural and medical biotechnology in the late 1990s served to undercut the lobbying clout of the biotechnology sector overall. It allowed the European Union to champion precautionary biosafety regulations on agricultural GMOs while protecting the interests of the pharmaceutical industry, which it sought to protect from being dragged further into the GM food controversy. Unlike in the other contentious areas of the biosafety talks, developing countries faced a united front of industrialized countries as both the US and the EU were determined to protect pharmaceutical biotech firms from new international regulations.

The last push for an international agreement, 1999–2000

In the changing international climate of 1999, all parties declared an interest in resolving the last remaining issues and reaching a compromise deal, in time for adoption at a resumed ExCOP meeting, scheduled to take place in January 2000 in Montreal. In order to prepare the ground for this, two consultative meetings were convened, in Montreal in July and in Vienna in September 1999. The Vienna meeting was conducted in a much calmer and friendlier atmosphere, and although producing no tangible agreement it signalled greater willingness on all sides to make concessions. The US delegation, whose assertive stance had been widely blamed for blocking an agreement at Cartagena, was now led by David Sandalow, who had stated in Congressional hearings that one of his priorities was to negotiate a biosafety treaty that protected US interests (International Environment Reporter, 1999c, 1999d, 1999e).

Based on the Vienna meeting, ExCOP chair Mayr produced a 'non-paper' in which he sought to distil the results of the consultative process into a compromise text that would guide subsequent negotiations. Mayr's summary identified the most difficult issues for the resumed ExCOP meeting that was scheduled to take place in Montreal in January 2000: the treatment of agricultural commodities, with regard to the application of the AIA procedure and identification requirements; the scope of the protocol; and trade-related issues, including the relationship of the protocol with other international regimes such as the WTO. This three-fold cluster of issues provided the basic structure for the negotiations in the final ExCOP meeting (Samper, 2002: 73).

With over 750 participants including delegates from 133 governments in attendance and anti-GMO campaigners protesting outside the conference centre, the Montreal ExCOP provided a potent reminder of how controversial the international biosafety talks had become by this time. Hopes had increased that the final round of negotiations would be successful, though many delegates gave the protocol only a 50–50 chance. Delegates made early progress in the negotiation group that focused on the scope of the protocol (Article 4). The agreement included an inclusive scope definition covering all GMOs, but with special provisions that would exempt certain GMOs altogether (e.g. pharmaceuticals) or exempt them from the protocol's main provisions. Once the chair of this group signalled substantial progress, the negotiations shifted to focus on the treatment of commodities and the trade-related issues in the protocol. The latter two issues were of paramount importance to the corporate lobbying groups at Montreal. The agricultural biotech industry and particularly the agricultural trading companies were keen to prevent the protocol from imposing unnecessarily cumbersome and costly regulations on GMO commodities.

If anything, it was the agricultural trading sector that was most adamantly opposed to the biosafety treaty and its treatment of the burgeoning trade in GMO commodities. Farm trade associations from Canada and the US argued strongly against the inclusion of commodities (living modified organisms for direct use as food, feed, or for processing, or LMO-FFPs, in the language of the protocol). They saw this as an unnecessary hurdle to trade and objected to the introduction of iden-tification requirements that would affect the entire GM and non-GM agricultural export sector. But their position was weakened by the fact that at the time of the biosafety talks, only a few national farming associations (mostly in North America) could be mobilized to lobby against the treaty. Most other major agricultural markets viewed the issue more from an importer perspective, with some farming sectors in Europe and elsewhere actively lobbying their governments to safeguard the future of non-GM farming (see below). The agricultural sector was thus an important ally of the biotechnology sector, but their global economic presence and thus political power was limited to North America and the few export-oriented countries of the Miami Group.

The presence of a large number of environment ministers in the last days of the meeting clearly underscored the political significance that many countries ascribed to the outcome of the talks. Step by step, negotiators were able to find compromise formulations that included a reference to precautionary decision-making under conditions of scientific uncertainty without directly referring to the precautionary principle. They also chose to fudge the issue of whether decisions taken under the protocol could be challenged with WTO law by inserting an ambiguous formulation in the preamble that declared trade and environmental agreements to be 'mutually supportive'. The one issue, however, that seemed to escape a compromise was the commercially most important question of how agricultural commodities, which accounted for the majority of international GMO trade, would be treated in the protocol (Falkner, 2002: 20–1).

Frantic last-minute negotiations involving only a small number of key delegates from the EU, the Like-Minded Group and the Miami Group (represented by Canada) sought to find a solution on this issue during the last night of the conference. By this time, the United States had decided that it did not want to see the biosafety talks collapse again and had come to accept a compromise proposal, according to which GMO shipments were to be accompanied by documentation simply stating that they 'may contain' GMOs (Article 18) without specifying which types of GMOs they included. Only Canada was now holding out against this formulation, which went beyond its negotiation mandate and was strongly opposed by Canadian agricultural exporters. Under pressure from all other parties including the US, Canada's environment minister David Anderson overruled his delegation's negotiation mandate on this issue, and the ExCOP plenary was able to adopt the final compromise text in the early hours of 29 January 2000.

The adoption of the Cartagena Protocol was a remarkable victory for those who had campaigned for this outcome since the early 1990s. Against opposition from a wide-ranging business front and North American governments, negotiators from the developing world and Europe secured a deal that, despite its many concessions to GMO exporters, allowed importing nations to carry out precautionary risk assessment of GMO trade. In the end, the GIC expressed its support for the agreement, though it was clear that many of its members remained critical about the new regulatory framework. GIC chairwoman Joyce Groote was quoted saying that the protocol 'will do what it was supposed to: protecting biodiversity without restricting trade', and that the compromise text recognized that biotechnology was an opportunity, not just a risk (BBC, 2000). The positive reaction could not hide the fact, however, that this agreement was achieved largely against business opposition. The strength of the global anti-GMO movement, combined with international leadership from developing countries and the EU, created the political momentum behind it, and divisions within the business sector undercut the anti-regulatory business lobby. The resolute stance of agricultural trading companies helped to water down key provisions on commodities trade, but could not stop the treaty from coming into existence. Business groups were able to shape the compromise text in important ways, but could not prevent it.

Beyond the Cartagena Protocol: regime evolution and the global governance of GM food

As is common in international negotiations, the Cartagena Protocol represented a diplomatic compromise that left some critical issues unresolved. Future negotiations among the parties would have to clarify certain treaty provisions and add those that were still missing. Before this could happen, however, the protocol needed to be ratified by the required number of parties, 50 in total, in order to enter into force. That the biosafety treaty would receive sufficient ratifications to become a binding international regime was not in doubt. The Like-Minded Group of developing countries that had been a driving force in the negotiations had sufficient members to reach the legally required quorum. The real question was whether any of the major industrialized countries, and particularly the main GMO-exporting nations, would ratify. For if the protocol was to become a comprehensive framework for global GMO governance, importer *and* exporter countries would need to become parties to it. The United States was unlikely to ratify any time soon, as it would need to first become a party to the CBD, the Cartagena Protocol's mother convention, and there was little hope that the US Senate would agree to this under the then Republican leadership. Canadian negotiators had signalled that they were willing to submit the treaty to parliament for ratification, but also made it clear that this depended on other parties taking Canadian concerns on board in further developing the treaty. Canada's farming and farm trade associations in particular voiced their opposition to early ratification until some of their key concerns had been addressed. It was therefore far from clear whether some of the major GMO-exporting nations would accede to the treaty.

Despite the uncertainty over the membership of the Cartagena Protocol, its norms and rules could still be expected to have at least a *de facto* impact on GMO trade. Even if some of the key biotechnology countries would not ratify the agreement, its provisions could still be enforced on them by importing countries that they sought to trade GMOs with. Article 24 of the protocol clarifies the relationship between parties and non-parties in international trade. It states that '[t]ransboundary movements of living modified organisms between Parties and non-Parties shall be consistent with the objective of this Protocol', and that 'Parties shall encourage non-Parties to adhere to this Protocol and to contribute appropriate information to the Biosafety Clearing-House on living modified organisms released in, or moved into or out of, areas within their national jurisdictions'. The formulation of this article leaves considerable room for manoeuvre, but most importing countries that have created domestic biosafety legislation in line with the Cartagena Protocol will not distinguish between different types of exporters in applying the domestic framework for risk assessment and management. As more and more major agricultural import markets are adopting Cartagena Protocol-style biosafety rules, a growing proportion of international GMO trade is therefore likely to be affected by the treaty's biosafety rules. This *de facto* reach of the regime is at least one important reason why major North American biotechnology and trading companies have continued to lobby the Meetings of the Parties to the Cartagena Protocol, even though their own governments have not acceded to the agreement.

The discussion so far has focused on the Cartagena Protocol as the centrepiece of global biosafety governance. But we should also consider the wider transnational context in which biotechnology and GMO trade are debated, negotiated, and to some extent governed. In this broader perspective, it becomes clear how business conflict – between biotech firms and the farming sector, and between GM food producers and food retailers – has shaped the evolution of agricultural biotechnology. Several sites of contestation can be identified in the global production network of GM food, which have provided anti-GMO campaigners with leverage over the production network and have allowed them to influence the commercialization of biotechnology in agricultural markets. Much of this has taken on the form of diffuse and irregular market governance, but has nevertheless had an important effect on how GM food has spread worldwide.

The following discussion first traces developments in the context of the Cartagena Protocol before broadening the perspective to consider the wider context of business conflict and private governance along the GM food production chain.

Revising and implementing the Cartagena Protocol

After the adoption of the protocol in January 2000, an Intergovernmental Committee for the Cartagena Protocol (ICCP) met three times to prepare for the first Meeting of the Parties. While the ICCP could not take binding decisions on the development of the protocol, it nevertheless helped to get the Biosafety Clearing House off the ground and made recommendations on a large list of outstanding issues. Industry groups were actively involved in the meetings, lobbying delegates in the corridors and hosting information sessions that highlighted the economic consequences of the various regulatory provisions that still needed to be finalized by the parties. In June 2003, the small island state of Palau submitted the fiftieth ratification of the protocol to the secretariat, and as per Article 37, the Cartagena Protocol entered into force 90 days later, in September 2003. The first meeting of the Conference of the Parties (COP) to the CBD serving as the Meeting of the Parties (MOP) to the Cartagena Protocol (COP/MOP-1) was convened to take place in February 2004 in Kuala Lumpur, Malaysia. Two further COP/MOP meetings were held subsequently, in Montreal, Canada, in May/June 2005 and in Curitiba, Brazil, in March 2006.

The main issues of contention in the first three COP/MOP meetings have been the identification of GMO content in agricultural commodity trade, the establishment of a compliance mechanism, and the further development of rules on capacity-building and liability (see Falkner and Gupta, 2004; Garton et al., 2006). The first issue has provoked considerable controversy as it directly affects the interests of agricultural trading companies, which have become the dominant force in business lobbying at the recent biosafety meetings. Their main interest has been to limit identification requirements to a minimum and to leave unchanged the protocol language in Article 18, which speaks of the requirement to identify GMO content in a shipment by stating in accompanying documentation that it 'may contain' GMOs. This formulation had been inserted into the draft treaty text at the last minute to secure its adoption in 2000, and the majority of parties were keen to replace it with a more definite formulation of 'contains' to give import authorities greater certainty about the content of such shipments. As before, the issue of clarifying this provision, in addition to the related questions of the threshold at which it kicks in, was at the centre of the conflict between exporters and importers at COP/MOP-1, 2 and 3, and was only resolved in 2006 by agreeing another compromise solution that distinguished between cases where the identity of the GMO in question is known or unknown through identity preservation systems, and left the language unchanged for the latter case.

Debates on the other critical issues on the implementation agenda have involved differences in positions more between developed and developing countries. On the design of the compliance mechanism, European countries argued for a stronger mechanism that would include the right to take sanctions against parties found to be in violation of protocol rules. This proposal came up against strong resistance from developing countries, who have traditionally favoured soft mechanisms that facilitate implementation by creating greater transparency and helping non-compliant countries with capacity-building. Likewise, the debate over a binding international liability regime, which would force producers and exporters of GMOs to compensate importer countries for any damage to the environment or public health from GMO releases, did not go very far amidst strong resistance from industrialized countries. As a compromise, the parties agreed to open a separate negotiation process on a liability regime, the success of which is far from guaranteed (Falkner and Gupta, 2004).

Two important changes, in the negotiation format and the global political economy of biotechnology, affected the political dynamics at the COP/MOP meetings. The first change concerned the rules of procedures that apply to meetings of the parties once a treaty has entered into force. Unlike in the negotiation phase that is open to all countries, only those countries that have ratified the agreement have the right to speak and vote on resolutions and treaty amendments. Since none of the major GMO-exporting countries had ratified the protocol, the first two COP/MOP meetings were dominated by GMO-importing nations. Other countries with an explicit exporter interest, such as the United States, Canada and Argentina, were in attendance as observer nations, but their participation in discussions was curtailed by the rules of procedure, especially when it came to critical debates in smaller contact group settings and decisions taken by COP/MOP. The parties were thus able to push for stronger decisions on how to fill the remaining gaps in the treaty, and the discussions in Kuala Lumpur and Montreal demonstrated the resolve of many developing countries and the EU to press ahead with the implementation process. However, if the major GMO exporters were to be encouraged to ratify the agreement, their views and concerns needed to be reflected, too. Canada, one of the leading GMO exporters that had announced its intention of acceding to the treaty, repeatedly reminded the parties that if they ignored the concerns of its agricultural trading sector, ratification of the treaty would become increasingly unlikely.

The second change of the negotiation dynamic arose from the slowly shifting landscape of global GMO commercialization. At the time of the Cartagena Protocol negotiations, only the United States, Canada and Argentina had significant commercial interests in GMO exports. By the time of the first COP/MOP meetings, however, a number of developing countries particularly in Latin America had joined the ranks of nations with commercial GM crop production, and their change in commercial outlook impacted on the positions they took in the international process. At COP/MOP-1 and 2, Brazil and Mexico emerged with a more explicit exporter interest. Brazil was already under domestic pressure in the run-up to the Montreal conference of 2000 to emphasize its agricultural export interest but remained attached to the Like-Minded Group (Nogueira, 2002). Since the signing of the Cartagena Protocol, GM soybeans have been planted first illegally in the country's southern regions and legally since 2003. The country saw a rapid takeup of the biotech varieties of its major export crop, and in 2006 Brazil became the world's third largest grower of GM crops, with a total area of GM planting of 11.5 million hectares (James, 2006). Mexico's decision to shift position had less to do with existing commercial interests in GM crops – it ranked 13th globally in 2006, with GM soybean and cotton grown on only 100,000 hectares (James, 2006) – but more with its membership in the North American Free Trade Agreement (NAFTA) and established agricultural trade links with the United States and Canada (Gupta and Falkner, 2006).

While Brazil retreated from its more explicit pro-trade stance at the third COP/MOP meeting in March 2006 – as the host nation, it had a strong interest in the talks reaching a compromise between the exporter and importer perspectives – other nations emerged as proponents of a more trade-friendly approach to developing the protocol's provisions. New Zealand, Paraguay and Peru came out strongly against the wishes of importing nations to create more stringent identification rules for GMOs in bulk commodity shipments (Garton et al., 2006). Their interventions were widely welcomed by the agricultural trading companies represented at the international meeting, who had felt for some time that the parties did not sufficiently acknowledge the reality of global commodities trade.

The shifts in country positions do not follow strictly, or even closely, underlying shifts in corporate interests, but the influence of shifting business interests cannot be ignored. Often, as in the case of Brazil, Mexico and New Zealand, political shifts are the result of internal wrangling between ministries that vie for control over international biosafety policy, especially between environment and trade ministries. The commercialization of GM crops remains controversial even in those countries that now represent a clear exporter perspective in international negotiations, such as Mexico and Brazil, and equally pronounced corporate interests are at play that seek to protect the export interests of non-GM farmers (Hochstetler, 2007). Still, the growing spread of GM crops in a number of agricultural markets has helped to shift the balance between pro-GMO and anti-GMO business interests in favour of the former. Concerns over technological innovation and competitiveness are increasingly playing into domestic battles over the future of biosafety governance, providing pro-GMO business interests increased legitimacy in domestic political debates.

Business conflict and global resistance to GM food

The Cartagena Protocol has now become the central element of the emerging global governance architecture for GMOs and GM food, though it is not the only such instrument. Other international treaties and institutions also provide governance functions that partly overlap, and to some extent clash, with the Cartagena Protocol's rules for precautionary risk regulation: the WTO's Agreement on the Application of Sanitary and Phytosanitary Measures (SPS Agreement) is the main trade-related regime for food safety and provides a more restrictive basis for taking precautionary trade measures; and the Codex Alimentarius Commission promotes the international coordination and harmonization of food standards, although it has so far failed to agree on a unified approach to GM food labelling (Isaac and Kerr, 2007). The relationship between these different elements of global GMO governance remains uncertain and contested, not least because the main protagonists in the global biosafety conflict, the US and the EU, disagree over the scope for precautionary regulation and the applicability of competing legal contexts in trade disputes (Drezner, 2007: 161–4; Falkner, 2007a). This became all too clear in the biosafety negotiations, and again in the WTO trade dispute that the US (together with Canada and Argentina) brought against the EU's GMO regulations in 2003. Although the WTO case did not directly challenge the Cartagena Protocol, it nevertheless brought out the fundamental disagreements between the two sides with regard to how WTO disciplines and precautionary biosafety regulation should be reconciled (Brack et al., 2003; Boisson de Chazournes and Mbengue, 2004).[3]

While the international governance architecture for GMOs remains fragmented and fractured, other forms of social steering and private governance have emerged outside the states system that have come to shape the direction of GMO commercialization and innovation. The activities of anti-biotech movements have succeeded not only in creating greater public awareness of the environmental and health risks associated with GM food but also in turning consumers away from the new technology. In extreme cases, they have led to the closure of entire markets to GM food products. This would not have been possible without the key role played by business power and business conflict. As Schurman argues, the anti-biotech movement 'strategically exploited certain key vulnerabilities of the industry' (2004: 244). These vulnerabilities derived from the nature of the GM food chain and the potential that exists for competition and conflict between different industries and firms along the chain. In this sense, business conflict has helped to empower social protest groups in the global struggle over GMO trade, and business power and private governance structures have been employed to direct the commercial evolution of agricultural biotechnology.

The private governance of GM food is indeed a good example of how business conflict can undermine the structural power of core business groups. The biotechnology industry comprises major multinational corporations that have developed genetic engineering into an innovative technology platform for revolutionary changes in a range of applications, from pharmaceuticals to agriculture and manufacturing. Once hailed as opening the doors to the 'biotech century' (Rifkin, 1998), genetic engineering has been supported by governments keen to capitalize on the technology as a new growth engine for the twenty-first century. Yet despite the rapid adoption of GM technology in agriculture following its commercial introduction in the mid-1990s, GM crops and food have been rejected in major agricultural markets and remain absent from supermarket shelves in Europe and in other countries. The industrial restructuring of the late 1990s, which ended the vision of an integrated life sciences sector, signalled the growing commercial and political risks that have befallen agri-biotechnology. While the major biotechnology companies have spent considerable resources on developing and marketing their innovations and have run multi-million dollar PR campaigns to win over farmers and consumers, social protest and business conflict along the GM food chain have severely dented the industry's prospects.

To illustrate the importance that business conflict plays as a lever for social movements and as a source of informal governance in global biotechnology, two cases are examined in more detail below, each located at a different point in the global GM food production chain. The first, at the consumer end of the chain, involves supermarkets that have effectively closed European consumer markets to GM food products. Their powerful economic position as large-scale food retailers and their focus on food safety and reputational risks has led them to control the presence and identification of GM content in food products. The second case of business conflict is located further down the chain towards the producer end, and involves farmers and commodity traders. Their concerns about the impact that new GM crops will have on export markets with GMO restrictions have become an important transmission mechanism for anti-biotech sentiment across national boundaries. In some cases, they have created new hurdles for the commercialization of GM crops and have forced biotechnology firms to establish consultative forums that assess the socio-economic consequences of new crops.

To begin with the first case of business conflict, once anti-GMO campaigners had targeted newly available GM food in the mid-1990s (Bauer et al., 2002), it was the food retail sector that came to play a critical role in the early demise of the new products. Supermarkets led the worldwide movement towards labelling GM food products, which many governments have come to mandate through legislation. In Europe, for example, supermarket chains introduced voluntary GM labels and even eliminated all GM content from their own-brand food products. Large supermarket chains used their market power over food manufacturers and suppliers to demand the identification, and in some cases elimination, of GM content in food production and distribution, which in turn sent strong signals further down the production chain to agricultural traders and farmers. By amplifying anti-GMO protests and consumer hostility, supermarkets thus raised the hurdles for broader market acceptance of GM food and shaped the market in a way that led to its closure for these novel products.

The changing political economy of food retailing in Europe explains why supermarkets were in such a strong position vis-à-vis biotech firms. The overwhelming trend in the European retail sector has been one of increasing levels of concentration, greater market share of own-label products selling under retailer names and greater coordination between retailers and suppliers along the supply chain. Large food retailers have built up a strong position of trust among consumers, which enhances their market strength but also makes them vulnerable to negative publicity and reputation-damaging controversies in the food safety area. On issues of critical importance to consumers, therefore, food retailers will exercise great caution and, if necessary, provide consumers with information on food ingredients through labelling schemes (Loader and Henson, 1998: 32–3).

The short-lived history of GM food in Europe started in 1996, when the first cans of GM tomato puree became commercially available. In February, two of the UK's largest retailers, J Sainsbury and Safeway, offered the GM variety next to non-GM products, clearly labelled as a GM product. As GM crops such as soybeans and corn were also being commercially grown in the United States from 1996 onwards, European retailers were keen to ensure that no new GM ingredients would find their way into the European food chain without the full knowledge of the retail sector. In early summer 1997, they therefore called upon US commodity suppliers to create segregated distribution channels for GM and non-GM food. Their demand was ignored, however, due to the additional costs this segregation would have imposed on US producers and traders. Faced with the first signs of unease among consumers and continuing uncertainty about the EU's legislative proposals for GM food labelling, British food retailers and manufacturers took the initiative and produced their own food-labelling code in November 1997. Whereas the EU's draft labelling scheme included references to food products that 'may contain' GM content, the UK retail sector's code opted for a definite form of labelling which indicates that a product 'contains' GM content (Nunn, 2000).

The retail sector did not stop at labelling. Iceland, a relatively small retailer with only 1.6 per cent of total UK grocery sales, announced in March 1998 that it would eliminate all GM ingredients from its own-label grocery products (Loader and Henson, 1998: 33). Sainsbury followed this move in July 1999 and became the first large British supermarket to claim that it had eliminated all GM content (mainly GM soya protein) from its own-brand products. The food retailer worked with over a thousand of its suppliers to ensure that only certified non-GM crops would enter the supply chain. Other major British supermarket chains, such as Tesco and Marks & Spencer, were also taking steps to eliminate GM foods from their shelves (Reuters, 1999). Under pressure from activist groups, leading supermarkets in Britain, Ireland, France and Italy then sought to coordinate their efforts and formed a consortium to increase their leverage vis-à-vis GM soya producers (Milmo, 1999). They all took voluntary actions to eliminate GM content from their food range and put pressure on their suppliers to do the same, or at a minimum to identify all GM food content in the production and distribution chain.

To some extent, European supermarkets acted pre-emptively, in anticipation of a European directive on GMO labelling (Novel Food Regulation), which had been delayed due to internal disagreements within EU institutions and came into force in 1997 (Loader and Henson,

1998: 33). But the main impetus for the introduction of voluntary GM labels was the fear that rising anti-GMO consumer sentiment would dent the retail sector's reputation for high levels of food safety. Ironically, opinion polls conducted before GM food labels began to appear in the UK showed no strong opinion on either side of the GM food debate (Loader and Henson, 1998: 31). Because concern over hostile consumer reaction was the retail sector's main concern, many large supermarkets went beyond existing regulatory requirements and implemented a complete programme of elimination of all GM content in own-label products. By 2005, 27 of the 30 top European retailers had adopted a policy of excluding GM ingredients from products in their European or main markets (Greenpeace, 2005).

By creating voluntary GM labelling schemes, eliminating GM ingredients from own-name products and even banning GM food products altogether, leading supermarkets hastened the demise of GM food in Europe. As this example shows, lack of unity between biotechnology firms, agricultural producers and food retailers has dealt the hopes for commercializing GM foods worldwide a severe blow. Due to their economic size and exposure to reputational risks, food retailers play a central role in relaying consumer preferences and societal values to food producers, agricultural traders and farmers. They may, of course, distort such preferences and values, seek to shape them, or merely amplify them. In any case, food retailers have a pervasive influence over agricultural practices further down the commodity chain, and this influence often extends across national boundaries into different farming sectors around the world. Food producers and traders, in turn, have responded to the influence of the retail sector by developing a strategy of market differentation and supply segregation, tailoring their products to the tastes and preferences of regional markets. In this way, business conflict between GM food producers and supermarkets has led to a segmented, and ultimately more limited, global market for GM food.

The second major form of business conflict in the GM food chain is between biotech firms on the one hand and farmers and agricultural traders on the other. While food retailers have transmitted anti-GMO consumer attitudes to food suppliers, it was farmers and traders that channelled these preferences to the biotech and seed sector. Indeed, one of the most important barriers to the adoption of new GM crop varieties has been resistance by the farming and agricultural trade community, in both developed and developing economies. There are several reasons why some farmers resist the adoption of GM crops: they may see the higher costs of GM seeds and the need to re-buy seeds every growing season as a deterrent; they may have doubts about the long-term economic benefits that can be derived from GM crops; they may fear negative consumer reactions to the sale of GM crops and GM food products, and thus the loss of market share particularly in export markets; and they may object to the growing dominance of a small number of biotech firms that are increasingly controlling the production and distribution of major commodity seeds.

Several instances of farmer and trader resistance to new GM crop varieties have been reported from around the world. In China, exporters of soybeans and soya-based products to Japan, Korea and the EU have lobbied the government against the start of commercial production of GM soybean varieties in the country, out of fear that co-mingling and accidental GM contamination of shipments abroad would close off foreign markets (Falkner, 2006). In India, rice exporters have protested against plans to test and introduce GM rice to Indian agriculture. In November 2006, the All-India Rice Exporters' Association called for an immediate end to all field trials of a GM rice variety developed by Monsanto-Mahyco, for fear of losing its position as the world's third largest exporter of basmati rice (Parsai, 2006). And in South Africa, the country's association of wine growers, the South Africa Wine Council, has rejected plans by biotech firms to introduce genetically modified yeast in wine production, also out of fear of losing valuable export markets (Cape Times, 2006).

These examples suggest that biotechnology firms have come up against considerable resistance in the global GM food chain. Most of these cases of business conflict have occurred outside North America, in countries where the success of agri-biotechnology has been more limited. However, global opposition to GM food has also begun to translate into business conflict in the North American GM food chain. A prominent example of this is the case of the failed introduction of GM wheat in the US and Canada. Having applied for regulatory approval of GM wheat in both countries, Monsanto was forced to withdraw the applications in 2004 and declared an end to research and development of GM wheat. While this withdrawal may be only of a temporary nature, the dramatic turnaround in Monsanto's strategy demonstrates the crucial role that conflicting corporate strategies between producers and users of GM seeds play in transmitting anti-GMO consumer attitudes.

This outcome is all the more remarkable given Monsanto's success with the introduction of GM soybeans and corn, which have captured around 90 and 50 per cent market share in the US respectively. The high-value 'hard red spring' variety of wheat, which is popular with flour mills because of its higher protein content, was identified as the next potential success story for genetic engineering, and Monsanto led the industry race in this field. In 2002, the company filed parallel applications to deregulate GM wheat in the US and Canada, and was also seeking regulatory approval in other countries, such as Japan and South Africa (Associated Press, 2004; Olson, 2005).

The regulatory applications encountered strong resistance from environmental campaigners and consumer groups, and most critically from wheat farmers and traders. Farming organizations in Canada and the United States expressed deep concerns about the commercial impact the new crop variety would have and voiced their opposition to regulatory approval for GM wheat. Their objections were based on weed management and economic cost concerns, but the dominant fear was that the introduction of GM wheat would threaten the export markets particularly in those countries with strict GMO import regimes. The growing rift between the biotechnology industry and wheat growers became all too clear when market studies pointed to persistent resistance to GM wheat in foreign markets, especially in Europe and Asia. In 2000–01, eight of the ten top importers of North American hard red spring wheat were Asian and European countries, and the overwhelming majority of buyers in Japan, Korea, China and the EU had publicly declared that they would not purchase GM wheat or shipments with GMO contamination, even if such varieties had passed the regulatory process in their markets (Olson, 2005: 157–8; Cropchoice, 2001). Economic studies of the short-term economic impact of GM wheat confirmed the fears of many farmers (Wisner, 2003).

The threat of losing export markets was further compounded by certain characteristics of global wheat production. Because wheat is grown primarily for food production and human consumption (unlike soybean and corn, which are predominantly processed into animal feed or intermediate products such as oils), it is much more susceptible to anti-GMO consumer attitudes. Also, American producers grow a much smaller share of global wheat production (8 per cent in 2002) and therefore face much stiffer international competition in export markets. If US wheat production were to switch to GM varieties, US wheat exports could easily be replaced with non-GM supplies from other producer countries. And finally, US wheat farmers have already experienced a long-term downward trend in their global market share, with the US share of wheat export down from nearly 50 per cent in the 1970s to just over 20 per cent in 2001 (Wisner, 2003). With downward pressure on prices for wheat in the US, farmers were particularly exposed to the threat of loss of export markets.

Recognizing the growing resistance among North American farmers, Monsanto set up a Wheat Industry Joint Biotech Committee in 2001 to seek a consensus with the farming community and assured farmers that a decision on commercialization would depend on market acceptance for the biotech crop (Fairchild, 2002). To be sure, not all wheat farmers were opposed to GM varieties, particularly those producing primarily for the domestic market. The National Association of Wheat Growers had been more sympathetic to Monsanto's developments while the US Wheat Associates voiced the concerns of exporters (Bernick and Wenzel, 2006). But in the end, divisions within the farming community and strong lobbying by exporter interests, who account for over half of the American wheat market (Wisner, 2004: 17), prevented a market consensus on GM wheat. Opposition to GM wheat commercialization ran particularly high among Canadian farmers, who are more dependent on export markets than their US counterparts (Johnston, 2003).

By 2004, it had become clear to Monsanto's executives that its consensus-building strategy had failed. Not wanting to further antagonize the farming community, the company announced in May that it was delaying plans to the commercial introduction of Roundup Ready wheat and that it would re-focus its R&D efforts to other crops such as corn, oilseeds and cotton (Burchett, 2004). Monsanto has since worked hard to persuade US farming associations to support in principle the commercial future for GM wheat (Bernick and Wenzel, 2006), but has had far less success with Canadian farming associations who remain opposed.

Monsanto's failure to introduce GM wheat in North America, the most receptive market environment for GM crops, is a prime example of the importance of business conflict to the future of biotechnology. While consumer resistance and environmental campaigns are the root cause of the industry's difficulties especially in food crops, corporations at the farm production and crop trading point in the GM food chain play a pivotal role in translating consumer preferences into market signals and barriers. The biotech industry's future success depends crucially on the willingness of farmers to adopt GM crop varieties, and in the case of wheat it was farmers that stalled the seemingly unstoppable progress of agri-biotechnology in North America.

Conclusions

When biosafety was first debated at the international level, in the run-up to the 1992 'Earth Summit', the first commercial applications of genetic engineering were still being developed in laboratories or tested in field trials. Agricultural biotechnology had grown rapidly in the preceding years to form the technological platform for a new industrial sector, but in the early 1990s the industry was still in its infancy. With the exception of the United States, which led the worldwide commercialization of genetically modified plants, agricultural biotechnology had not yet reached the status of a major industrial sector, although the promise of genetic engineering to revolutionize agriculture and other economic sectors had endeared it to governments keen to nurture what they perceived as the growth engines of the future.

When the possibility of regulating biotechnology internationally was first discussed in the context of the negotiations on the Biodiversity Convention, it was biotechnology firms particularly in the US that fought the creation of such an international agenda. From then on, the biotechnology industry led the business campaign against international biosafety regulations, and was joined only later by agricultural interest groups once GM crops had been commercialized. Throughout the 1990s, North American biotech firms were more actively involved in international lobbying than their European counterparts, reflecting the different levels of biotechnological commercialization and industry consolidation on both sides of the Atlantic. But despite these regional differences, the agricultural biotechnology sector remained united in its stance throughout the international negotiations, largely due to the growing internationalization of the industry.

At around the time when international negotiations on a biosafety protocol to the CBD started, GM crops started appearing in agriculture, first in North American and then in international trade. The commercialization of genetically modified crops in the second half of the 1990s mobilized agricultural farming and trading interests who gradually became more involved in the international process. The arrival of a greater number of agricultural lobbyists, who opposed the imposition of new biosafety rules on commodities trade, allowed the biotechnology sector to broaden its lobbying campaign and gain an important ally. Supported by the US and Canada, the anti-regulatory alliance of biotech firms and farming organizations succeeded in weakening some of the key trade-relevant provisions in the biosafety treaty, but were unable to prevent it from being adopted.

In the end, business conflict severely constrained the strength of the anti-regulatory lobby. North American agricultural trading interests found few international allies for their campaign against the emerging biosafety regime, as most other agricultural sectors in Europe and the developing world viewed the issues of biosafety from an importer perspective.

Strong anti-GMO campaigns and hostile consumer reactions in Europe and elsewhere had curtailed the growth of agri-biotechnology outside North America, thus shifting the balance of power between anti- and pro-regulatory forces in favour of the latter. The international divide in the Cartagena Protocol negotiations, which pitted the EU and a large group of developing countries against a small US-led group of agricultural exporter countries, reflected therefore the underlying political economy of biotechnology.

Business conflict, and particularly the failure to create a global business front in support of agricultural biotechnology, helped to pave the way for the adoption of the biosafety treaty in 2000. For most of the international negotiations, the biotechnology sector was able to maintain a uniform international stance, despite transatlantic differences in lobbying style and strategy. Their anti-regulatory case was weakened more by the fact that biotechnology was still a nascent and fragmented industrial sector in many industrialized countries. While hostile consumer reactions in Europe and elsewhere had prevented the emergence of a broader coalition of biotech and agricultural interests on a global scale, they also led to growing divisions within the biotechnology sector. By the late 1990s, the agricultural and pharmaceutical biotechnology sector increasingly went their own commercial ways, which in turn further weakened the lobbying stance of the biotech firms in the end phase of the biosafety talks. Agricultural biotechnology never managed to develop the kind of structural power that the fossil fuel industry possesses in climate politics.

Part III: Conclusions

6
International Environmental Politics and Business Power: Conclusions and Implications

Business involvement in international environmental politics: opportunities and threats

The three cases examined in this book – ozone layer depletion, climate change and agricultural biotechnology – reveal an unambiguous trend towards greater business involvement in international environmental politics. As the global environmental agenda has steadily grown since the first United Nations environment conference in 1972, business actors have developed a greater and more organized presence at the international level. Global firms in particular now consider international environmental rules and regulations to be an integral part of their global political strategy, and new global business associations have sprung up that deal exclusively with environmental policy issues (e.g. WBCSD, GCC, ICCP). Few companies will follow all international processes associated with the now over 200 MEAs, but sectoral business associations for major industries (e.g. chemicals, automobiles, biotechnology) routinely engage in those international debates and negotiations that matter most to their membership base.

The growing involvement of business in international environmental politics has been documented in other issue areas, too. International efforts to create norms and rules for international trade in hazardous waste (Clapp, 2001), ocean fisheries (DeSombre, 2000), timber trade and forestry management (Gale, 1998), non-timber forest products (Shanley et al., 2002) and persistent organic pollutants (Clapp, 2003), to name but a few, have all attracted a broad range of corporations and other nonstate actors. As the scope of international environmental policy-making has expanded and affected ever more markets, companies have been slowly drawn in to the global governance architecture that has emerged around these issues. Moreover, many global firms now routinely engage in global environmental debates as part of their overall commitment to corporate social responsibility (Vogel, 2005).

This should come as no surprise, for with the growth of global environmental governance, the specific roles of companies – as producers, traders, innovators and investors – have taken on a more eminently political dimension. To be effective, global environmental standards and rules need to be translated into economic incentives and constraints that operate in national and global markets. In this, the operations of companies, their decisions on product and process development, on marketing, innovation and competitive strategy, all play an important role that can decide whether environmental regulations manage to bring about the required socio-economic change. In this sense, corporations provide critical governance functions in the global economy, be it as part of formal international environmental regimes or informal private governance mechanisms.

This trend is the inevitable outcome of the enmeshing of global environmental and economic agendas. It has long been recognized that the expansion of the environmental agenda has brought international environmental regulation into closer contact with other policy domains. On the one hand, this has led to greater friction between different regime contexts, as can be observed in the tensions between international environmental agreements and WTO rules (Eckersley, 2004). On the other hand, policy integration across different issue areas is necessary to deal with the complex problems that global issues such as climate change or the regulation of biosafety pose. The increase in inter-regime interaction can therefore lead to both conflict and opportunities for synergy effects (Oberthür and Gehring, 2006), and the same can be said for the merging of global environmental and corporate agendas.

Policy-makers have responded to these challenges by creating more formal mechanisms for business involvement that seek to manage this problematic interface. Apart from setting the rules of nonstate actor participation in international environmental negotiations (on climate change, see Depledge, 2005), business representatives have been invited into institutional mechanisms that support environmental treaties and provide expert advice to decision-makers. The creation of technology assessment panels in the Montreal Protocol has set a successful example in this area, allowing industry representatives to provide expert advice and thus to shape regulatory discourses particularly on questions of economic impacts and technological uncertainty. In the climate regime, business representatives have also offered advice on mitigation strategies and novel technologies, and have come to play a critical role in the flexibility instruments of the protocol such as the CDM and emissions trading. Furthermore, in a number of private governance mechanisms, businesses and NGOs have come to create monitoring and certification schemes to support compliance with private standards. To be sure, the involvement of business actors as experts in environmental governance arouses suspicions of 'regulatory capture'. It has certainly enhanced the legitimacy of the corporate sector as a partner in the search for international solutions and provides business with new levers of influence that lobbying alone would not achieve. However, it is also a sign of the maturing relationship between states, environmental NGOs and business organizations, and the growing recognition that complex global problems require the support and coordination of a wide range of actors.

While there has been an overall, long-term, trend towards greater international involvement by the business sector, corporate participation in specific international agendas and political processes follows a more cyclical pattern. As can be seen in the three cases examined in this book, most companies usually respond to emerging environmental issues in a reactive and obstructive way, often failing to anticipate the long-term challenges that these may pose for their business operations. They routinely challenge the scientific basis of perceived new environmental threats and warn against precautionary measures, but often find it difficult to develop proactive environmental strategies at an early stage. As soon as environmental issues become the focus of international negotiations, business actors begin to develop a clearer understanding of the problem and how international action might impact on them, but given the often unpredictable and drawn-out nature of environmental diplomacy, only the most directly affected and globally active companies will be involved from the start of negotiations. Over time, as the diplomatic process develops focus and momentum, business involvement will grow, and business lobbyists will follow more closely the negotiations in order to shape the outcome. Once an international agreement has been reached and signatories begin to implement it at the domestic level, the full range of affected business interests will be drawn in to the policy debate. It is often at the domestic implementation stage that all relevant business sectors will mobilize to shape the way international rules are translated into the domestic context, or to block ratification of an international accord altogether. At this point in the international policy cycle, business actors will have been able to assess more clearly the costs and benefits of international regulation, and are likely to exploit most effectively their established links with regulators and legislatures in the domestic context.

The growing involvement of business therefore presents both opportunities and threats to environmental policy-makers and campaigners. Given their critical role as polluters and providers of technological solutions, constructive engagement with the business sector is inevitable for effective environmental governance. The many cooperative arrangements that have sprung up in recent years, between states and firms, NGOs and firms, and among firms themselves, all indicate the opportunities that have opened up for leveraging business support in the pursuit of environmental sustainability. At the same time, the mobilization of business in international environmental policy-making has led to shifts in the discourse on regulatory politics, with greater emphasis on business- and market-friendly measures that have challenged traditional concepts of environmental governance. Some of these new instruments may strengthen global governance overall, but have also led to contestation of state authority (Conca, 2005) and established accountability norms (Koenig-Archibugi, 2004).

Business power and conflict in ozone, climate and biosafety politics

The three cases examined in this book demonstrate how business engagement with the global environmental agenda has deepened since the 1980s. Over time, as global environmental issues have moved from agenda setting, treaty negotiation, regime evolution and implementation, an ever greater number of business actors have become involved in the international process. They have developed more precisely defined corporate interests and strategies with regard to international regulation, created new lobbying organizations to represent their case and formed new political alliances across the spectrum of international actors to further their interests.

Despite this general trend towards ever greater involvement of business, there have been noticeable imbalances in business representation in the cases examined above. In ozone layer politics, the producers of CFCs, the chemical substances responsible for the depletion of stratospheric ozone, led the business response from the beginning and dominated international business lobbying during the Montreal Protocol negotiations and beyond. They were the most immediately affected business sector by the growing ozone controversy, and they possessed the resources and experience to engage in international negotiations. By contrast, many of the CFC user industries came late to the international process and relied heavily on the chemical industry to fight global CFC restrictions. They were thus caught off-guard when DuPont and other chemical firms changed their strategy and came out in support of an international CFC phase-out programme. After the Montreal Protocol was agreed in 1987, more and more user industries woke up to the challenge of the regulatory changes and became more active in the international process, signalling a gradual end to the predominance of the handful of major CFC producers.

The early phase of international climate politics saw the creation of a formidable alliance of corporate actors that were threatened most directly by proposed restrictions on greenhouse gas emissions: the fossil fuel industry, comprising oil and coal companies and major industrial manufacturing sectors that depended on fossil fuels as energy or industrial inputs. Led by the oil multinationals, the fossil fuel industry became the dominant business lobby group in the 1992 'Earth Summit' negotiations on the framework convention and in the run-up to the 1997 Kyoto Protocol talks. Given the wide range of economic and social activities implicated in global warming, other business sectors also became involved in the international debate, though their presence in the international process never came to rival that of the fossil fuel industry's main lobbying organizations, the Global Climate Coalition and the Climate Council. Other business interests were lacking in economic strength (e.g. renewable energy firms) or failed to develop an effective and sustained political strategy (e.g. insurance industry). In the end, it was the disintegration of the fossil fuel lobby in the early to mid-1990s that significantly increased diversity in business representation and lobbying strategies, with new groupings such as the International Climate Change Partnership and the World Business Council for Sustainable Development taking a more conciliatory stance.

In the international politics of biosafety, the biotechnology industry led the business campaign against international regulations from the start of the international debate in the early 1990s. North American biotech firms were more actively involved in international lobbying than their European counterparts, reflecting the different levels of bio-technological commercialization and industry consolidation on both sides of the Atlantic. Despite these regional differences, the agricultural biotechnology sector remained united in its stance throughout the international negotiations, largely due to the growing internationalization of the industry. As genetically modified crops came to be commercialized and started to permeate international farm trade in the second half of the 1990s, agricultural farming and trading interests became more involved in the international process. The arrival of a greater number of agricultural lobbyists, who opposed the imposition of new biosafety rules on commodities trade, allowed the biotechnology sector to broaden its lobbying campaign. In the end, it was North American agricultural trading interests that remained most strongly opposed to the emerging biosafety regime, long after the biotechnology industry had lent its support to the Cartagena Protocol.

The evolution of business lobbying in these three issue areas demonstrates how business representation has expanded while becoming more diverse as the international process moves from agenda setting to negotiation and implementation. International environmental regulation creates differential effects on business. As the number of politically engaged business actors has expanded, so has the possibility for business conflict. The international politics of ozone depletion, climate change and agri-biotechnology show how companies and business associations have not pulled in the same direction, and how divisions and conflict between different business interests have characterized the political role of business.

Business conflict has had important political consequences. In all cases, it has undermined business power overall, and particularly in the case of anti-regulatory forces that tend to dominate business lobbying in the early phases. But whether latent divisions in the business sector would develop into business conflict, and whether such conflict would significantly change international political dynamics, depended on the relative strength of competing business interests and contingent factors of the international process. In ozone politics, the strategic divisions between different CFC producers played a key role in regime creation. They promoted an international consensus on the need to take precautionary CFC restrictions, and later divisions between CFC producers and users helped to speed up the strengthening of the Montreal Protocol's CFC phase-out programme. Given how central the small group of CFC producers was to the problem of ozone layer depletion, business conflict among them was bound to reverberate through the negotiation process.

In climate change, the divisions between the fossil fuel industry and the business interests that stood to gain from international regulation were present from an early stage but did not translate directly into a politically significant business conflict, largely due to underlying imbalances in economic strength and political capacity. The later decline of business unity in the fossil fuel industry – between energy providers and industrial manufacturers, and within the oil industry itself – came to play a more important political role in shaping the political strategies of the major players in climate politics during the 1990s, the EU and the US. The growing diversity of business interests and strategies since the adoption of the Kyoto Protocol has created new opportunities for progressive political alliances in favour of climate action and governance, though the underlying structural centrality of fossil fuel-based energy has restricted the overall impact of business conflict in the politics of climate change.

In the early phase of biosafety politics, the biotechnology sector was able to maintain a uniform international stance, despite transatlantic differences in lobbying style and strategy. Their anti-regulatory case was weakened more by the fact that biotechnology was still a nascent and fragmented industrial sector in many industrialized countries. The arrival of genetically modified crops in international agriculture helped to mobilize agricultural trading interests that joined biotechnology firms in their fight against biosafety regulation. But GM crops were only beginning to be commercialized in a small number of countries at a time when the international biosafety negotiations were already under way. Furthermore, hostile consumer reactions in Europe and elsewhere prevented a broader coalition of biotech and agricultural interests on a global scale, and the anti-regulatory coalition of biotech and agricultural interests was therefore mostly confined to North America. Their lobbying position was further weakened by growing divisions within the biotechnology sector, which in the late 1990s led to the disintegration of the life sciences industry into separate agricultural and pharmaceutical biotech sectors.

One major effect of greater business involvement has been that it has created new hurdles in the search for global environmental solutions. As is common in interest group politics at every level, those interests that are most likely to be negatively affected by proposals for new regulations are the ones most likely to seek to influence, and if possible prevent, their creation. Oil companies campaigned most strongly against binding commitments for greenhouse gas emission reductions; the chemical industry fought plans to reduce the production of ozone-depleting chemicals; and the biotechnology industry opposed the creation of international biosafety rules. Their lobbying efforts have slowed down the international process, and while they could not prevent the creation of international regulations they have had a significant influence over their design.

At the same time, the deepening engagement of business with international environmental politics has also led to a greater diversity of business interests being represented internationally. As proposals to create environmental regulations gathered momentum and cracks in the initial business front began to appear, pro-regulatory business interests have emerged that created political space for political leaders to push for stricter environmental standards. Greater business involvement has therefore had a more mixed effect on the creation of global environmental governance than the high-profile campaigns of anti-regulatory business groups would suggest. Business groups can now be found on both sides of the environmental debate. In this way, the new political alliances that business divisions and conflict have given rise to, have contributed to the pluralization of the international politics of the environment.

Implications for the study of business power

The case studies in this book underline the importance of ongoing efforts in International Relations that raise awareness about the international political role of business (e.g. Fuchs, 2005; May, 2006). This book has shown how in recent years business actors have become increasingly engaged in international environmental politics; how they have shaped international political processes and the outcomes of regime-building efforts; and how they have interacted with a range of international actors – states, international organizations and environmental campaign groups – to create governance structures for environment, both within and outside the international states system. Business actors matter, therefore, in environmental politics as in other international policy areas, because they constrain or widen policy options, influence the formation of interests and shape regulatory discourses, and provide governance functions within the global economy.

The field of international environmental politics provides important lessons for wider debates on how to think about business power and its limits. In seeking to influence international outcomes, business actors rely on multiple dimensions of power: relational, structural and discursive. Relational power, the ability to prevail over other actors in situations of conflict, has been clearly visible in the environmental field, wherever business actors have lobbied governments and sought to influence the design of international regimes. Overall, the business sector possesses superior financial resources and strong organizational capacity, particularly when compared to environmental NGOs, and is well placed to exploit the privileged access it has to key governmental actors. However, these power resources have not always translated into a predominant position in international environmental politics, and have been challenged by civil society groups' ability to overcome financial constraints through transnational networking and coordination. The key role that environmental ministries play in MEA negotiations has also deprived business of the advantage of close working relationships with more business-friendly government officials. The issue-specific characteristics of environmental negotiations and the rise of a new and imaginative form of transnational activism have thus served to curtail the business sector's relational power.

Our analysis would be too limited if we did not also take into account the business sector's central position in the global economy, which gives rise to structural business power. It is usually in this area that business is credited with a dominant, even privileged, position as it controls decisions on investment and technological innovation. The case studies in this book have shown how this dimension of power plays into the dynamics of international environmental politics. Corporations possess structural power in the traditional sense, in that policy-makers need to consider the broader economic impact that proposed environmental regulation will have. They also possess what can be described as technological power, in that corporations largely shape perceptions of which policy options are technologically, and by implication economically, feasible. In this sense, corporations indirectly shape international outcomes, by setting parameters for policy-makers. But the analysis in this book also suggests that structural power needs to be translated into the international process through the agency of firms, and that we need to consider the contingent ways in which business actors bring structural power to bear. Divisions among them greatly limit the sector's overall structural power, and have in many cases opened up opportunities to overcome structural barriers through political agency. Likewise, the discursive power of the business sectors has been undermined by a lack of business unity and been challenged by environmental campaign groups that question the legitimacy of business actors.

The neo-pluralist perspective advanced in this book not only urges us to study business power in its empirical manifestations within issue-specific contexts, but also draws our attention to the close connections that exist between business power and business conflict. As can be seen in international environmental politics, inter-firm and inter-sectoral conflict is always a latent reality, and frequently serves to limit business power overall. Whether business conflict manifests itself and comes to shape business involvement in international politics depends on several factors, including the nature of the issue at hand, industry structures and the effects of regulatory politics. It is also influenced by the agency of other actors who seek to exploit the political opportunities of business conflict. Political pressure and social protest play an important role in creating the conditions for business conflict to emerge.

The business conflict model holds important lessons for political leaders and civil society actors that seek to steer society and the economy in the direction of greater environmental sustainability. It suggests that the dynamics of economic competition and the potential for conflict between corporations may enhance the capacity of campaign groups to exert pressure on companies and bring about a change in corporate behaviour. Social movement theorists speak of 'industry opportunity structures' (Schurman, 2004) that empower activist groups in their political campaigns. Where the potential for business conflict exists, for example between market leaders and laggards, or between companies operating at different points in transnational production chains, activist groups have sought to exploit these divisions and create political alliances with companies more likely to support stricter international environmental standards. For example, NGOs such as Greenpeace have sought to enlist the support of the insurance industry and other financial sectors to push for precautionary measures against the threat of global warming (Leggett, 1996).

Environmental activist groups have traditionally targeted states and international organizations in order to promote international norms and rules that bind economic actors and force change upon them. While this remains an important avenue of influence for NGOs, social movements have long come to realize that it is not the only, or even most promising, strategic option available to them. A growing number of activists have engaged in what Wapner (1996) calls 'world civic politics', which involves targeting multinational corporations directly and creating global governance structures outside the states system. Here again, business conflict provides activists with access points and powerful levers that allow them to pressure companies into change. It opens opportunities for such groups to engage and cooperate with more progressive companies in an effort to change markets and establish norms for good corporate behaviour. For example, environmental NGOs have campaigned for supermarkets and other food retailers to eliminate genetically modified food from their product range, which has in turn sent strong market signals to other businesses along the international food chain, including farmers and agricultural trading companies. In ozone politics, Greenpeace went as far as teaming up with a refrigeration company in Germany to market a novel refrigeration system (Greenfreeze) in the hope of changing market dynamics and forcing other companies to adopt the technology.

As Cerny has observed, political globalization that accompanies global economic integration has resulted in a situation where outcomes 'are determined not by simple coercion and/or structural power but, even more significantly, by how coalitions and networks are built in real time conditions among a plurality of actors' (2003: 156). Indeed, the proliferation of political alliances between diverse sets of actors, involving states, NGOs and business actors, makes for a more pluralistic and open-ended international politics of the environment. It does not create a level playing field for competition among equals. Significant power imbalances persist, and structural business power can constrain the search for global political solutions to environmental problems. But business does not determine outcomes in international environmental politics, nor can it control the global environmental agenda.

Notes

Chapter 1   Global firms in international environmental politics

1. For stylistic reasons, EU is used throughout the book to refer to the European Community before, and the European Union after, entry into force of the Maastricht Treaty in 1993.

Chapter 3   Ozone layer depletion

1. Estimates of the workforce involved in the CFC business varied greatly. At the beginning of the ozone controversy, DuPont claimed that only 200,000 workers in the US were involved in CFC-related operations (US House of Representatives, 1975: 377). The later 1976 estimate of 1 million workers was subsequently revised by a US Commerce Department analysis which concluded that only about 594,000 of these workers truly represented CFC-dependent employment, most of which (83.3 per cent) were in refrigeration and air conditioning, 11.4 per cent in foams and plastics, 4.6 per cent in aerosols, and 0.7 per cent in precursor chemical and CFC production (Gladwin et al., 1982: 79).
2. Reporting of ozone-related issues in chemical industry trade journals fell significantly during the early 1980s. *European Chemical News*, one of the leading magazines in the field, failed to even mention the Vienna Convention in its coverage of regulatory events in 1985.
3. This section draws on Falkner, 2005.
4. The Montreal Protocol did, of course, create an uneven incentive structure for phasing out ODS in that it did not cover all ozone-depleting chemicals as regulated substances from the outset. By covering only five CFCs (11, 12, 113, 114, 115) and three halons (1211, 1301, 2402) – a compromise reflecting the delicate balance between precautionary action and commercial interests – the negotiators of the Montreal Protocol created a regulatory framework that allowed users of unregulated substances (e.g. HCFCs, methyl chloroform, carbon tetrachloride) to delay action for many years. But within the group of regulated substances, the protocol did not differentiate between different usage types.

Chapter 4   Global climate change

1. Climate Change Initiatives and Programs in the States, Arkansas through Montana. Pew Center on Global Climate Change, February 2007. Available at: <http://www.pewclimate.org/docUploads/States%20table%204.26.pdf>.

Chapter 5   Agricultural biotechnology

1. The Cartagena Protocol on Biosafety uses the term 'living modified organism' (LMO), but this chapter uses the more common term 'genetically modified organism' (GMO).
2. The EU's final negotiation mandate for the 2000 Montreal conference was agreed by the EU Environment Ministers on 13 December 1999, and was made public in a press release on 27 January 2000 (2235th Council Meeting – Environment, Brussels, 13/14 December 1999. Conseil/99/409).
3. The EU lost the WTO case in the first instance in 2006. As the Cartagena Protocol was not the subject of the US case against the EU, the WTO panel did not rule on the relationship between the protocol and WTO rules.

References

Aerosol Age (1975a) 'Who Dropped the Ball?' May, pp. 53–4.
—— (1975b) 'Allied's Orfeo Stresses Uncertainties in Ozone Theory', June, pp. 38–43.
—— (1975c) 'Changing Public Opinion', June, pp. 16–18.
—— (1975d) 'Business Strategies for the Future', November, pp. 18–23.
—— (1975e) 'The Search for Alternatives', November, pp. 28–31.
—— (1975f) 'The Aerosol Market in Japan', December, pp. 44–50, 71.
—— (1976a) 'The Changing Worldwide Market', January, pp. 16–17.
—— (1976b) 'The Ozone Issue – Who's in Control?' July, pp. 26–8.
—— (1976c) 'Hydrocarbons Gain in Market Share', July, pp. 22–4.
—— (1976d) 'Alternate Fluorocarbon Propellants – 142b and 22', December, pp. 45–6, 58.
—— (1977) 'Use 11 and 12 For Now... But Consider the Alternatives', January, pp. 32, 39.
—— (1978) 'British Production Sets Record in 1977', July, pp. 20–1.
Air Conditioning, Heating & Refrigeration News (1987a) 'Refrigerant Squeeze Will Hit Wholesalers Hard', 20 April, p. 3.
—— (1987b) 'Uncertain CFC Status Clouds Residential Surge', 15 June, p. 6.
—— (1989) 'Government Understands Industry's Needs on CFCs', 24 April, p. 28.
Alliance for Responsible CFC Policy (1986a) 'A Search for Alternatives to the Current Commercial Chlorofluorocarbons', submitted to UNEP Economic Workshop on Protection of the Ozone Layer, May 1986, Washington, DC (dated 24 February).
—— (1986b) 'Economic Consequences of the United States Ban on the Use of Fully-Halogenated Chlorofluorocarbons as Aerosol Propellants', Washington, DC.
—— (1989) 'HCFCs and HFCs Provide the Balance', Washington, DC.
Akhurst, M., J. Morgheim and R. Lewis (2003) 'Greenhouse Gas Emissions Trading in BP', *Energy Policy* 31(7): 657–63.
Andrée, P. (2007) *Genetically-Modified Diplomacy: The Global Politics of Agricultural Biotechnology and the Environment* (Vancouver: University of British Columbia Press).
Andrews, D.M. (1994) 'Capital Mobility and State Autonomy: Toward a Structural Theory of International Monetary Relations', *International Studies Quarterly* 38(2): 193–218.
Arts, B. (1998) *The Political Influence of Global NGOs: Case Studies on the Climate and Biodiversity Conventions* (Utrecht: International Books).
Associated Press (2004) 'Monsanto Plans for Biotech Wheat On Hold', 10 May.
Aulisi, A., A.E. Farrell, J. Pershing and S. VanDeveer (2005) *Greenhouse Gas Emissions Trading in U.S. States: Observations and Lessons from the OTC NO_x Budget Program.* WRI White Paper. Washington, DC: World Resources Institute.

Automotive Environment Analyst (1998) 'Business Support for CO_2 Emissions Trading', 1 June.

Ayres, E., and H. French (1996) 'The Refrigerator Revolution', *World Watch* September/October: 14–21.

Bachrach, P., and M.S. Baratz (1970) *Power and Poverty: Theory and Practice* (New York: Oxford University Press).

Bail, C., J.P. Decaestecker and M. Jørgensen (2002) 'The European Union', in C. Bail, R. Falkner and H. Marquard (eds) *The Cartagena Protocol on Biosafety: Reconciling Trade in Biotechnology with Environment and Development?* (London: Earthscan), pp. 166–85.

Bail, C., R. Falkner and H. Marquard (eds) (2002) *The Cartagena Protocol on Biosafety: Reconciling Trade in Biotechnology with Environment and Development?* (London: Earthscan).

Ballhorn, R. (2002) 'Canada', in C. Bail, R. Falkner and H. Marquard (eds) *The Cartagena Protocol on Biosafety: Reconciling Trade in Biotechnology with Environment and Development?* (London: Earthscan), pp. 105–14.

Barnet, R., and J. Cavanagh (1994) *Global Dreams: Imperial Corporations and the New World Order* (New York: Simon & Schuster).

Barnett, M., and R. Duvall (2005) 'Power in International Politics', *International Organization* 59(1): 39–75.

Baron, D.P. (2006) *Business and Its Environment*, 5th edn (Upper Saddle River, NJ: Pearson Prentice Hall).

Barringer, F. (2006) 'Officials Reach California Deal to Cut Emissions', *New York Times*, 13 August.

Bauer, M.W., G. Gaskell and J. Durant (eds) (2002) *Biotechnology: The Making of a Global Controversy* (Cambridge: Cambridge University Press).

Baumert, K.A., T. Herzog and J. Pershing (2005) *Navigating the Numbers: Greenhouse Gas Data and International Climate Policy* (Washington, DC: World Resources Institute).

BBC News (2000) 'GM Deal Finds Favour All Round', 29 January.

Beard, C.A. (1934) *The Idea of National Interest: An Analytical Study in American Foreign Policy* (New York: The Macmillan Company).

Benedick, R.E. (1991) *Ozone Diplomacy: New Directions in Safeguarding the Planet* (Cambridge, MA: Harvard University Press).

Bennhold, K. (2004) 'EU Companies Coming Under Kyoto Limits: A New Era in Battling Pollution', *International Herald Tribune*, 31 December.

—— (2005) 'Global Heat on Bush Increases', *International Herald Tribune*, 28 January.

Bernauer, T. (2003) *Genes, Trade, and Regulation: The Seeds of Conflict in Food Biotechnology* (Princeton, NJ: Princeton University Press).

Bernick, J., and W. Wenzel (2006) 'Finally, a Biotech Statement from the Wheat Organizations', *Farm Journal*, 8 March.

Bernstein, S. (2001) *The Compromise of Liberal Environmentalism* (New York: Columbia University Press).

Betsill, M.M., and H. Bulkeley (2004) 'Transnational Networks and Global Environmental Governance: The Cities for Climate Protection Network', *International Studies Quarterly* 48(2): 471–93.

Bio/Technology (1991) 'Editorial: Canary in a Coal Mine', 9(4): 313.

Birnie, P., and A. Boyle (2002) *International Law and the Environment*, 2nd edn (Oxford: Oxford University Press).

Boddewyn, J.J., and T.L. Brewer (1994) 'International-Business Political Behavior: New Theoretical Directions', *Academy of Management Review* 19(1): 119–43.

Boisson de Chazournes, L., and M.M. Mbengue (2004) 'GMOs and Trade: Issues at Stake in the EC Biotech Dispute', *Review of European Community and International Environmental Law* 13(3): 289–305.

Bomann-Larsen, L., and O. Wiggen (eds) (2004) *Responsibility in World Business: Managing Harmful Side-effects of Corporate Activity* (Tokyo: United Nations University Press).

Børsting, G., and G. Fermann (1997) 'Climate Change Turning Political: Conference-Diplomacy and Institution-Building to Rio and Beyond', in G. Fermann (ed.) *International Politics of Climate Change: Key Issues and Critical Actors* (Oslo: Scandinavian University Press), pp. 53–82.

Boulton, L. (1997) 'Oil Chief Presses Case for Solar Power', *Financial Times*, 20 May.

—— (1998) 'Business Groups to Lobby MPs on Climate Change', *Financial Times*, 23 March.

Boville, B.W. (1979) 'Environmental Aspects of Stratospheric Ozone Depletion', in A.K. Biswas (ed.) *The Ozone Layer: Proceedings of the Meeting of Experts Designated by Government, Intergovernmental and Nongovernmental Organizations on the Ozone Layer Organized by the United Nations Environment Programme in Washington DC, 1–9 March 1977* (Oxford: Pergamon Press).

Brack, D., R. Falkner and J. Goll (2003) 'The Next Trade War? GM Products, the Cartagena Protocol and the WTO', Briefing Paper No. 8, London: Royal Institute of International Affairs.

Brewer, T.L. (2005) 'Business Perspectives on the EU Emissions Trading Scheme', *Climate Policy* 5(1): 137–44.

Brickman, R., S. Jasanoff and T. Ilgen (1985) *Controlling Chemicals: The Politics of Regulation in Europe and the United States* (Ithaca, NY: Cornell University Press).

Brown, P., and J. Vidal (1999) 'GM Investors Told to Sell Their Share', *Guardian*, 25 August.

Buchner, B., and S. Dall'Olio (2005) 'Russia and the Kyoto Protocol: The Long Road to Ratification', *Transition Studies Review* 12(2): 349–82.

Burchett, A. (2004) 'Monsanto Mothballs Roundup Ready Wheat', AGWEB, 10 May.

Business Week (1980) 'Ending Washington's Feud with Business', 24 November, p. 156.

Business Wire (1997) 'CAPP Gravely Concerned About Kyoto Agreement', 12 December.

Callon, M. (ed.) (1998) *The Laws of the Markets* (Oxford: Blackwell).

Cantley, M.F. (1995) 'The Regulation of Modern Biotechnology: A Historical and European Perspective', in D. Brauer (ed.) *Biotechnology: A Multi-volume Comprehensive Treatise. Vol. 12: Legal, Economic and Ethical Dimensions* (Weinheim: VCH), pp. 505–681.

Cape Times (2006) 'GM Firms' Wine Industry's Attempt Deflated', 27 October.

Carpenter, C. (2001) 'Businesses, Green Groups and the Media: The Role of Non-Governmental Organizations in the Climate Change Debate', *International Affairs* 77(2): 313–28.

Cerny, P.G. (2003) 'The Uneven Pluralization of World Politics', in A. Hülsemeyer (ed.) *Globalization in the Twenty-First Century: Convergence or Divergence?* (Basingstoke: Palgrave Macmillan), pp. 153–75.

Chemical and Engineering News (1987a) 'New Questions Posed in Antarctic Ozone Loss', 1 June, p. 6.

—— (1987b) 'Complex Mission Set to Probe Origins of Antarctic Ozone Hole', 17 August, pp. 7–13.

—— (1988) 'Search Intensifies for Alternatives to Ozone-Depleting Halocarbons', 8 February, p. 18.

Chemical Engineering (1988a) 'Chemical Firms Search for Ozone-Saving Compounds', 18 January.

—— (1988b) 'Firms Ally to Speed Development of CFCs', 25 April.

Chemical Manufacturers Association (1991) 'Production, Sales and Calculated Release of CFC-11 and CFC-12 Through 1989' (Washington, DC: CMA).

Chemical Week (1975) 'Fluorocarbons Canned', 25 June, p. 18.

Christiansen, A.C., and J. Wettestad (2003) 'The EU as a Frontrunner on Greenhouse Gas Emissions Trading: How did it happen and will the EU succeed?' *Climate Policy* 3(1): 3–18.

Clapp, J. (2001) *Toxic Exports: The Transfer of Hazardous Wastes from Rich to Poor Countries* (Ithaca, NY: Cornell University Press).

—— (2003) 'Transnational Corporate Interests and Global Environmental Governance: Negotiating Rules for Agricultural Biotechnology and Chemicals', *Environmental Politics* 12(4): 1–23.

—— (2006) 'Unplanned Exposure to Genetically Modified Organisms: Divergent Responses in the Global South', *Journal of Environment and Development* 15(1): 3–21.

—— (2007) 'Transnational Corporate Interests in International Biosafety Negotiations', in R. Falkner (ed.) *The International Politics of Genetically Modified Food: Diplomacy, Trade and Law* (Basingstoke: Palgrave Macmillan), pp. 34–47.

Cogan, D.G. (2006) *Corporate Governance and Climate Change: Making the Connection* (Boston, MA: Ceres).

Commission of the European Communities (1988) 'The Greenhouse Effect and the Community', Communication to the Council. Commission Work Programme Concerning the Evaluation of Policy Options to Deal with the 'Greenhouse Effect'. COM (88) 656 final, 16 November (Brussels: CEC).

—— (1991) 'A Community Strategy to Limit Carbon Dioxide Emissions and to Improve Energy Efficiency'. Communication from the Commission to the Council. SEC (91) 1744 final, 14 October (Brussels: CEC).

—— (1999) 'Preparing for Implementation of the Kyoto Protocol', Communication from the European Commission to the Council and the Parliament, COM(1999) 230, 19 May (Brussels: CEC).

—— (2000) 'Green Paper on Greenhouse Gas Emissions Trading Within the European Union', COM(2000) 87 final, 8 March (Brussels: CEC).

Conca, K. (2005) 'Old States in New Bottles? The Hybridization of Authority in Global Environmental Governance', in J. Barry and R. Eckersley (eds) *The State and the Global Ecological Crisis* (Cambridge, MA: MIT Press), pp. 181–205.

Cook, E. (1996) *Marking a Milestone in Ozone Protection: Learning from the CFC Phase-Out* (Washington, DC: World Resources Institute).

Cox, R.W. (1994) *Power and Profits: U.S. Policy in Central America* (Lexington: University Press of Kentucky).

—— (ed.) (1996) *Business and the State in International Relations* (Boulder, CO: Westview Press).

Cropchoice (2001) 'Global Opposition Mounts Against Monsanto's "FrankenWheat"', 1 February.

Cutler, A.C., V. Haufler and T. Porter (eds) (1999) *Private Authority and International Affairs* (Albany: State University of New York Press).

Daneshkhu, S., and F. Harvey (2006) 'Multinationals Nudged into Action', *Financial Times*, 3 February.

Dauvergne, P. (2001) *Loggers and Degradation in the Asia Pacific: Corporations and Environmental Management* (Cambridge: Cambridge University Press).

Davenport, D.S. (2005) 'An Alternative Explanation for the Failure of the UNCED Forest Negotiations', *Global Environmental Politics* 5(1): 105–30.

Davoudi, S. (2006) 'Monsanto Strengthens Its Grip on GM Market', *Financial Times*, 16 November.

Depledge, J. (1999) 'Coming of Age at Buenos Aires: The Climate Change Regime After Kyoto', *Environment* 41(7): 15–20.

—— (2005) *The Organization of Global Negotiations: Constructing the Climate Change Regime* (London: Earthscan).

Der Spiegel (1988) 'Es geht darum, unsere Haut zu retten', 24 October.

Der Stern (1989) 'Dicke Gewinne mit Ozonkillern', 18 May.

DeSombre, E.R. (2000) *Domestic Sources of International Environmental Policy: Industry, Environmentalists, and U.S. Power* (Cambridge, MA: MIT Press).

Dibner, M.D., G.N. Stock and N.P. Greis (1992) 'Away From Home: U.S. Sites of European and Japanese Biotech R&D', *Bio/Technology* 10(12): 1535–8.

Dicken, P. (2003) *Global Shift: Transforming the World Economy*, 3rd edn (London: Paul Chapman).

DiMaggio, P.J., and W.W. Powell (eds) (1991) *The New Institutionalism in Organizational Analysis* (Chicago: University of Chicago Press).

Doniger, D.D. (1988) 'Politics of the Ozone Layer', *Issues in Science and Technology* 4(3): 86–92.

Donnelly, J. (2007) 'Debate Over Global Warming is Shifting', *Boston Globe*, 15 February.

Doremus, P.N., W.W. Keller, L.W. Pauly and S. Reich (1998) *The Myth of the Global Corporation* (Princeton, NJ: Princeton University Press).

Dorey, E. (1999) 'EuropaBio Unit Created to Boost Agbio Defense', *Nature Biotechnology* 17(7): 631–2.

Drezner, D.W. (2007) *All Politics is Global: Explaining International Regulatory Regimes* (Princeton, NJ: Princeton University Press).

Dunleavy, P., and B. O'Leary (1987) *Theories of the State: The Politics of Liberal Democracy* (London: Macmillan).

Dunn, S. (1995) 'The Berlin Climate Change Summit: Implications for International Environmental Law', *International Environment Reporter* 18(11): 439–44.

—— (2002) 'Down to Business on Climate Change: An Overview of Corporate Strategies', *Greener Management International* (39): 27–41.

—— (2005) 'Down to Business on Climate Change: An Overview of Corporate Strategies', in K. Begg, F. Van der Woerd and D. Levy (eds) *The Business of Climate Change: Corporate Responses to Kyoto* (Sheffield: Greenleaf), pp. 31–46.

DuPont, Inc. (1979) 'Position Paper: Chlorofluorocarbon/Ozone Depletion Issue', Wilmington, DE: DuPont, August.

Dür, A., and D. De Bièvre (2007) 'The Question of Interest Group Influence', *Journal of Public Policy* 27(1): 1–12.

Earth Negotiations Bulletin (1999) 'Summary of the Fifth Conference of the Parties to the Framework Convention on Climate Change, 25 October – 5 November 1999', 12(123), 8 November.

Eckersley, R. (ed.) (1995) *Markets, the State and the Environment: Towards Integration* (Basingstoke: Palgrave Macmillan).

—— (2004) 'The Big Chill: The WTO and Multilateral Environmental Agreements', *Global Environmental Politics* 4(2): 24–50.

The Economist (1992) 'Europe's Industries Play Dirty', 9 May, pp. 91–2.

Eden, S.E. (1994) 'Using Sustainable Development: The Business Case', *Global Environmental Change* 4(2): 160–7.

Egziabher, T.B.G. (2002) 'Ethiopia', in C. Bail, R. Falkner and H. Marquard (eds) *The Cartagena Protocol on Biosafety: Reconciling Trade in Biotechnology with Environment and Development?* (London: Earthscan), pp. 115–23.

Eichhammer, W., et al. (2001) 'Greenhouse Gas Reductions in Germany and the UK – Coincidence or Policy Induced? An analysis for international climate policy', Research Report 20141133, UBA-FB 000193, Berlin: Umweltbundesamt.

Elliott, L. (2004) *The Global Politics of the Environment*, 2nd edn (London: Macmillan).

ENDS Report (1987) 'Firms Push Substitutes for Ozone-Depleting Aerosols', 152, September: 4.

—— (1989) 'Majority of Electronics Firms Lack Plans to Replace CFCs', 168, January: 5–6.

Enright, C.A. (2002) 'United States', in C. Bail, R. Falkner and H. Marquard (eds) *The Cartagena Protocol on Biosafety: Reconciling Trade in Biotechnology with Environment and Development?* (London: Earthscan), pp. 95–104.

Environmental Protection Agency (1997) 'Champions of the World: Stratospheric Ozone Protection Awards', EPA430-R-97–023, Washington, DC: Environmental Protection Agency, August.

Europa-Chemie (1987a) 'Rücknahme von FCKW-Kältemitteln', 19 March, p. 113.

—— (1987b) 'VCR bekräftigt Verzicht', 15 July, p. 327.

—— (1987c) 'Aerosol-Firmen setzen Reduzierung fort', 24 August, p. 377.

European Chemical News (1975) 'Fluorocarbon Impact Research Stepped Up', 2 May, p. 8.

—— (1978) 'Fluorocarbon "Reductions Without Regulations" Aim', 15 December, p. 20.

—— (1980a) 'EPA to Limit Fluorocarbon Production in US', 28 April, p. 14.

—— (1980b) 'Du Pont Increases Fluorocarbon Capacity', 22/29 December, p. 28.

—— (1981) 'Ozone Depletion by CFCs Less than Once Feared', 26 January, p. 19.

—— (1985) 'Firms Keep Their Nerve in Bhopal Aftermath', 4 March, pp. 12–3.

—— (1987a) 'CFC Lobby Planned', 23 March, p. 22.

—— (1987b) 'CFC Makers Face Severe Cutbacks in Production', 30 March, p. 6.

—— (1987c) 'Atochem Criticizes Terms of Global CFC Agreement', 28 September, p. 5.

—— (1989) 'ICS-Kali in CFC Link, Du Pont Patents Blends', 6 February, p. 17.

Evans, P.B., H.K. Jacobson and R.D. Putnam (eds) (1993) *Double-Edged Diplomacy: International Bargaining and Domestic Politics* (Berkeley: University of California Press).

Fairchild, B. (2002) 'Release Date for GM Wheat is 2005', *Farm Journal*, April.

Falk, R. (1997) 'State of Siege: Will Globalization Win Out?' *International Affairs* 73(1): 123–36.

Falkner, R. (1998) 'The Multilateral Ozone Fund of the Montreal Protocol', *Global Environmental Change* 8(2): 171–5.

—— (2001) 'Business Conflict and U.S. International Environmental Policy: Ozone, Climate, and Biodiversity', in P.G. Harris (ed.) *The Environment, International Relations, and U.S. Foreign Policy* (Washington, DC: Georgetown University Press), pp. 157–77.

—— (2002) 'Negotiating the Biosafety Protocol: The International Process', in C. Bail, R. Falkner and H. Marquard (eds) *The Cartagena Protocol on Biosafety: Reconciling Trade in Biotechnology with Environment and Development?* (London: RIIA/Earthscan), pp. 3–22.

—— (2003) 'Private Environmental Governance and International Relations: Exploring the Links', *Global Environmental Politics* 3(2): 72–87.

—— (2005) 'The Business of Ozone Layer Protection: Corporate Power in Regime Evolution', in D.L. Levy and P.J. Newell (eds) *The Business of Global Environmental Governance* (Cambridge, MA: MIT Press), pp. 105–34.

—— (2006) 'International Sources of Environmental Policy Change in China: The Case of Genetically Modified Food', *The Pacific Review* 19(4): 473–94.

—— (2007a) 'International Cooperation Against the Hegemon: The Cartagena Protocol on Biosafety', in R. Falkner (ed.) *The International Politics of Genetically Modified Food: Diplomacy, Trade and Law* (Basingstoke: Palgrave Macmillan), pp. 15–33.

—— (2007b) 'The Political Economy of "Normative Power" Europe: EU Environmental Leadership in International Biotechnology Regulation', *Journal of European Public Policy* 14(4): 507–26.

Falkner, R., and A. Gupta (2004) 'Implementing the Biosafety Protocol: Key Challenges', *SDP Briefing Paper 04/04*, London: Chatham House.

Farman, J.C., B.G. Gardiner and J.D. Shanklin (1985) 'Large Losses of Total Ozone in Antarctica Reveal Seasonal ClO_x/NO_x Interaction', *Nature* (315): 207–10.

Fligstein, N. (1990) *The Transformation of Corporate Control* (Cambridge, MA: Harvard University Press).

—— (2002) *The Architecture of Markets: An Economic Sociology of Twenty-First-Century Capitalist Societies* (Princeton, NJ: Princeton University Press).

Florini, A. (ed.) (2000) *The Third Force: The Rise of Transnational Civil Society* (Washington, DC: Carnegie Endowment for International Peace).

Fox, J.L. (1991) 'Quayle Likes Biotech, Not Regulation', *Bio/Technology* 9(4): 322–5.

—— (1999) 'Anti-GM Crop Protesters Increase Activity in the US', *Nature Biotechnology* 17(11): 1053–4.

Frankfurter Allgemeine Zeitung (1987a) 'Chemie will auf schädliche Spray-Gase verzichten', 22 May.

—— (1987b) 'Die Spraydose great unter stärkeren Druck', 18 August.

—— (1988) 'Umweltfreundlicherer Bau von Kühlschränken', 26 September.

—— (1989) 'Ein Schiff voller Demonstranten', 6 July.

Frieden, J. (1988) 'Sectoral Conflict and U.S. Foreign Economic Policy, 1914–1940', *International Organization* 42(1): 59–90.

Fuchs, D. (2005) *Understanding Business Power in Global Governance* (Baden-Baden: Nomos).

Fulton, M., and K. Giannakas (2001) 'Agricultural Biotechnology and Industry Structure', *AgBioForum* 4(2): 137–51.

Gale, F.P. (1998) *The Tropical Timber Trade Regime* (New York: St Martin's Press).

Garcia-Johnson, R. (2000) *Exporting Environmentalism: U.S. Multinational Chemical Corporations in Brazil and Mexico* (Cambridge, MA: MIT Press).

Garrett, G. (1998) *Partisan Politics in the Global Economy* (Cambridge: Cambridge University Press).

Garton, A., with R. Falkner and R.G. Tarasofsky (2006) 'Documentation Requirements Under the Cartagena Protocol on Biosafety: The Decision by the 3rd Meeting of Parties on Article 18.2(a)', Background note, London: Chatham House, 4 July.

Gelbspan, R. (1997) *The Heat Is On: The High Stakes Battle Over Earth's Threatened Climate* (Reading, MA: Addison Wesley).

Gereffi, G. (1994) 'The Organization of Buyer-Driven Global Commodity Chains: How U.S. Retailers Shape Overseas Production Networks', in G. Gereffi and M. Korzeniewicz (eds) *Commodity Chains and Global Capitalism* (Westport, CT: Praeger), pp. 95–122.

Gibbs, D.N. (1991) *The Political Economy of Third World Intervention: Mines, Money, and U.S. Policy in the Congo Crisis* (Chicago: University of Chicago Press).

Gilpin, R. (1975) *U.S. Power and the Multinational Corporation: The Political Economy of Foreign Direct Investment* (New York: Basic Books).

—— (1981) *War and Change in World Politics* (Cambridge: Cambridge University Press).

Giorgetti, C. (1999) 'From Rio to Kyoto: A Study of the Involvement of Non-Governmental Organizations in the Negotiations on Climate Change', *New York University Environmental Law Journal* 7(2): 201–45.

Gladwin, T.N., J.L. Ugelow and I. Walter (1982) 'A Global View of CFC Sources and Policies to Reduce Emissions', in J.H. Cumberland, J.R. Hibbs and I. Hoch (eds) *The Economics of Managing Chlorofluorocarbons: Stratospheric Ozone and Climate Issues* (Washington, DC: Resources for the Future), pp. 64–113.

Glas, J.P. (1988) 'Du Pont's Position on CFCs', *Forum for Applied Research and Public Policy* 3(3): 71–2.

—— (1989) 'Company Policy in the Face of Global Concerns: The Ozone Issue as a Model', Wilmington: Du Pont Company.

Glatzer, M., and D. Rueschemeyer (eds) (2005) *Globalization and the Future of the Welfare State* (Pittsburgh: University of Pittsburgh Press).

Gore, A. (2006) *An Inconvenient Truth: The Planetary Emergency of Global Warming and What We Can Do About It* (Emmaus, PA: Rodale Books).

Gow, D. (2005) 'CO_2 Trading Targets Too Generous, Say Environmentalists', *Guardian*, 5 January.

Grabner, P., J. Hampel, N. Lindsey and H. Torgersen (2001) 'Biopolitical Diversity: The Challenge of Multilevel Policy-Making', in G. Gaskell and M.W. Bauer (eds) *Biotechnology 1996–2000: The Years of Controversy* (London: Science Museum), pp. 15–34.

Grant, J. (2004) 'CCX Takes Chicago By Storm', *Financial Times*, 16 November.

Greenpeace (1996) 'Radically Accelerated Phase Out of Methyl Bromide and HCFCs with Controls on HFCs: An Environmental Imperative', Greenpeace International Position Paper, November.

—— (2005) 'No Market for GM Labelled Food in Europe', Greenpeace International, January.

Greenwood, J., and K. Ronit (1995) 'European Bioindustry', in J. Greenwood (ed.) *European Casebook on Business Alliances* (London: Prentice Hall), pp. 75–85.

Grubb, M., with D. Brack and C. Vrolijk (1999) *The Kyoto Protocol: A Guide and Assessment* (London: Earthscan).

Grubb, M., and F. Yamin (2001) 'Climatic Collapse at The Hague: What Happened, Why, and Where Do We Go From Here?' *International Affairs* 77(2): 261–76.

Gupta, A. (2000) 'Governing Trade in Genetically Modified Organisms: The Cartagena Protocol on Biosafety', *Environment* 42(4): 23–33.

Gupta, A., and R. Falkner (2006) 'Implementing the Cartagena Protocol: Comparing Mexico, China and South Africa', *Global Environmental Politics* 6(4): 23–55.

Guzzini, S. (1993) 'Structural Power: The Limits of Neorealist Power Analysis', *International Organization* 47(3): 443–78.

—— (2005) 'The Concept of Power: A Constructivist Analysis', *Millennium* 33(3): 495–522.

Haas, P.M. (1992) 'Banning Chlorofluorocarbons: Epistemic Community Efforts to Protect Stratospheric Ozone', *International Organization* 46(1): 187–224.

Haas, P.M., M.A. Levy and E.A. Parson (1992) 'Appraising the Earth Summit: How Should We Judge UNCED's Success?' *Environment* 34(8): 6–33.

Hajer, M.A. (1995) *The Politics of Environmental Discourse: Ecological Modernization and the Policy Process* (Oxford: Clarendon Press).

Hatch, M. (1993) 'Domestic Politics and International Negotiations: The Politics of Global Warming in the United States', *Journal of Environment and Development* 2(2): 1–39.

Haufler, V. (1995) 'Crossing the Boundaries between Public and Private: International Regimes and Non-State Actors', in V. Rittberger (ed.) *Regime Theory and International Relations* (Oxford: Clarendon Press), pp. 94–111.

Hay, C. (2002) *Political Analysis* (Basingstoke: Palgrave Macmillan).

Held, D., A. McGrew, D. Goldblatt and J. Perraton (1999) *Global Transformations: Politics, Economics and Culture* (Cambridge: Polity).

Henderson, J., P. Dicken, M. Hess, N. Coe and H.W.-C. Yeung (2002) 'Global Production Networks and the Analysis of Economic Development', *Review of International Political Economy* 9(3): 436–64.

Hertz, N. (2001) *The Silent Takeover: Global Capitalism and the Death of Democracy* (London: Heinemann).

Hildyard, N. (1993) 'Foxes in Charge of the Chickens', in W. Sachs (ed.) *Global Ecology: A New Arena of Political Conflict* (London: Zed Books), pp. 22–35.

Hileman, B. (1997) 'Storm Warning Rattle Insurers', *Chemical and Engineering News*, 14 April, pp. 28–31.

Hillman, A., and G. Keim (1995) 'International Variation in the Business–Government Interface: Institutional and Organizational Considerations', *Academy of Management Review* 20(1): 193–214.

Hillman, A.J., and M.A. Hitt (1999) 'Corporate Political Strategy Formulation: A Model of Approach, Participation, and Strategy Decisions', *Academy of Management Review* 24(4): 825–42.

Hirst, P., and G. Thompson (2000) *Globalization in Question: The International Economy and the Possibilities of Governance*, 2nd edn (Cambridge: Polity).

Hochstetler, K. (2007) 'The Multilevel Governance of GM Food in Mercosur', in R. Falkner (ed.) *The International Politics of Genetically Modified Food: Diplomacy, Trade and Law* (Basingstoke: Palgrave Macmillan), pp. 157–73.

Hodgson, J. (1990a) 'European Advisory Group Bares its Teeth', *Bio/Technology* 8(3): 185.

—— (1990b) 'Growing Plants and Growing Companies', *Bio/Technology* 8(7): 624–8.

—— (1991) 'EC Policy: Harmonize or Compromise?' *Bio/Technology* 9(6): 504.

—— (1992) 'Europe, Maastricht, and Biotechnology', *Bio/Technology* 10(11): 1421–6.

—— (2000) 'Moratorium Hits Danish Companies', *Nature Biotechnology* 18(2): 139–40.

Hoffman, A.J. (1997) *From Heresy to Dogma: An Institutional History of Corporate Environmentalism* (San Francisco: The New Lexington Press).

—— (2005) 'Climate Change Strategy: The Business Logic Behind Voluntary Greenhouse Gas Reductions', *California Management Review* 47(3): 21–46.

—— (2006) 'Getting Ahead of the Curve: Corporate Strategies that Address Climate Change', prepared for the Pew Center on Global Climate Change, University of Michigan, October.

Hopgood, S. (1998) *American Foreign Environmental Policy and the Power of the State* (Oxford: Oxford University Press).

Houghton, J. (2004) *Global Warming: The Complete Briefing*, 3rd edn (Cambridge: Cambridge University Press).

Houlder, V. (1998) 'Business Grapples with Climate Change', *Financial Times*, 11 November.

Hounshell, D.A., and J.K. Smith, Jr. (1988) *Science and Corporate Strategy: Du Pont R&D, 1902–1980* (Cambridge: Cambridge University Press).

Hoyle, R. (1992) 'Deep-Sixing Biodiversity', *Bio/Technology* 10(8): 848.

Humphreys, D. (2001) 'Forest Negotiations at the United Nations: Explaining Cooperation and Discord', *Forest Policy and Economics* 3(2): 125–35.

Hurrell, A., and B. Kingsbury (1992) 'The International Politics of the Environment: An Introduction', in A. Hurrell and B. Kingsbury (eds) *The International Politics of the Environment: Actors, Interests, and Institutions* (Oxford: Clarendon Press), pp. 1–47.

IAL Consultants Ltd (1986) 'Possible Effects of Additional Control Measures on CFC's in the U.K. Aerosol Industry', IAL Consultants Ltd.

IFC Inc. (1986) 'An Analysis of the Economic Effects of Regulatory and Non-Regulatory Events Related to the Abandonment of Chlorofluorocarbons as Aerosol Propellants in the United States from 1970 to 1980, With a Discussion of Applicability of the Analysis to Other Nations', Washington, DC: IFC Inc.

Ikwue, T., and J. Skea (1994) 'Business and the Genesis of the European Community Carbon Tax Proposal', *Business Strategy and the Environment* 3(2): 1–11.

International Environment Reporter (1979a) 'Munich Conference on Fluorocarbons Told Problems Demand World Attention', 10 January, pp. 477–8.

—— (1979b) 'Canadian Proposed Chlorofluorocarbon Regulations', 13 June, pp. 738–9.

—— (1979c) 'Proposed EEC Council Decision Concerning Chlorofluorocarbons in the Environment', 13 June, pp. 740–1.

—— (1979d) 'Non-Aerosol Spray Cans to be Labeled with "Environmental Friendliness" Seal', 11 July, p. 767.

—— (1980a) 'Rule on Chlorofluorocarbon Cap Expected in June, EPA Official Says', 14 May, p. 170.

—— (1980b) 'EEC Issues Chlorofluorocarbon Report Questioning Validity of 1979 NASA Study', 10 September, p. 401.

—— (1982) 'Divisions Develop at Stockholm Meeting as Group Drafts Ozone Protection Plan', 14 April, p. 142.

—— (1983a) 'Nations Still Unable to Reach Accord on Convention to Protect Ozone Layer', 12 January, pp. 9–10.

—— (1983b) 'Convention on Protecting Ozone Layer Nears Agreement, with Compromise on CFCs', 9 November, p. 503.

—— (1983c) 'Industry Opposes Worldwide CFC Ban, Questions U.S. Position at UNEP Meeting', 14 December, pp. 548–9.

—— (1984a) 'U.S. to Back Global Ban on Some CFC Uses over Industry's Objections, Officials Say', 11 January, pp. 7–8.

—— (1984b) 'UNEP Governing Council Forges Compromise on Convention, Protocol to Regulate CFCs', 13 June, pp. 181–2.

—— (1984c) 'Several Options to Control CFC Emissions Developed for Protocol to Ozone Convention', 14 November, p. 345.

—— (1986) 'Some UNEP Workshop Delegates Say Protocol on Ozone Protection Possible by Spring 1987', 8 October, pp. 346–7.

—— (1987) 'French Support Freeze, Not Reduction, of CFCs, Calling Evidence Inconclusive', 13 May, pp. 198–9.

—— (1988) 'Companies Should Plan on Bigger Cuts in CFC Production, Consumption, NRDC Says', 10 February, p. 111.

—— (1989) 'EC Council Agrees on Total Ban of CFCs by 2000; Bush Says US Goal Dovetails', March, p. 105.

—— (1992a) 'Estimates of Global Warming May be Exaggerated, IPCC Says', 29 January, p. 35.

—— (1992b) 'Global Warming Treaty Work Will Be Mostly Refining Language, Chairman Says', 26 February, pp. 91–3.

—— (1992c) 'Environment Commissioner Faults Bush on Refusal to Commit to Emission Limits', 8 April, pp. 184–5.

—— (1992d) 'Japan May Change Mind on CO_2 Reductions; Considers Backing U.S. at U.N. Conference', 22 April, p. 222.

—— (1992e) 'Nations Criticize New Draft Treaty as too Vague on Actions to be Taken', 6 May, pp. 253–4.

—— (1992f) 'Industry Officials Call Biodiversity Treaty Vague, See Little Impact on Competitiveness, Other Areas', 29 July, p. 512.

—— (1993a) 'Ministers Approve Stepped Up Timetable to Phase Out Ozone Depleting Substances', 13 January, p. 7.

—— (1993b) 'EPA Proposes Listing Methyl Bromide as Ozone Depleter Slated for Phase Out', 24 March, p. 209.

—— (1993c) 'EEC Proposal Sets Earlier Phase-out Date for HCFCs than Montreal Protocol Requires', 16 June, p. 428.

—— (1993d) 'Automakers Seek Options to Accommodate Phase-out of Coolants that Deplete Ozone', 30 June, p. 500.

—— (1993e) 'Propane/Butane, HFCs, Other Chemicals Being Tried by German Refrigerator Makers', 14 July, pp. 525–6.

—— (1994a) 'Environmentalists Criticize Request by United States for More CFC Production', 12 January, pp. 22–3.

—— (1994b) 'U.S. Official Says Emissions Commitments Under Climate Change Treaty Not Adequate', 9 February, pp. 112–13.

—— (1994c) 'Chairman of Climate Change Committee Claims Progress, but Industry, NGOs Disagree', 23 February, pp. 155–6.

—— (1994d) 'Hoechst Becomes First Chemical Company to Stop Production of Chlorofluorocarbons', 4 May, p. 390.

—— (1994e) 'Climate Change Uncertainties Mean Abatement Measures Unjustified, Report Says', 15 June, p. 515.

—— (1994f) 'Industry, Environmental Groups Asked to Give Policy Advice to State Department', 29 June, p. 576.

—— (1995a) 'Industry Says Climate Change Talks Should Focus on Existing Treaty Provisions', 8 February, p. 93.

—— (1995b) 'Ruling Government Attacked for Failure to Gain Support for Climate Treaty Protocol', 22 March, p. 217.

—— (1995c) 'U.K. Calls on Developed Countries to Set New Target for Greenhouse Gas Cuts by 2010', 22 March, pp. 217–18.

—— (1995d) 'Commission Announces New Policy Options to Meet Climate Change Treaty Requirements', 8 March, pp. 169–70.

—— (1995e) 'Concrete Action on Protocol Deferred; Two-Year Negotiation Process Established', 19 April, pp. 283–4.

—— (1995f) 'Coalition Protests Joint Implementation, Emissions Plan Drawn at Climate Conference', 19 April, pp. 284–5.

—— (1995g) 'Senate Would Reject Any Protocol Imposing Economic Hardship, Conference Told', 20 September, p. 720.

—— (1995h) 'Montreal Protocol Parties Should Weigh Risks, Not Just Costs, Greenpeace Says', 29 November, pp. 903–4.

—— (1996a) 'Group Gives U.S. President Recommendations on Curbing Vehicle Greenhouse Gas Emissions', 10 January, p. 22.

—— (1996b) 'Memorandum on Energy Use Targets 33 Percent Improvement in Efficiency by 2020', 24 January, pp. 61–2.

—— (1996c) 'Consideration of Joint Implementation Recommended by Environment Commissioner', 21 February, pp. 118–19.

—— (1996d) 'Talks Progress, Optimism Expressed for New Objectives Under Treaty by Fall 1997', 20 March, pp. 215–16.

—— (1996e) 'Energy Council Takes Preemptive Strike Against Forthcoming IPCC Recommendations', 1 May, pp. 361–2.

—— (1996f) 'Investment Cycles for Businesses Should be Considered by Policy-Makers', 26 June, p. 558.

—— (1996g) 'COP-2 Opens Facing Calls for More Action, Widening Gap Between Industry, Green Groups', 10 July, pp. 587–8.

—— (1996h) 'Biosafety Working Group Outlines Elements for Future Biodiversity Convention Protocol', 7 August, pp. 688–9.

—— (1996i) 'Green Groups, Industry Continue to Battle over Trade in Genetically Modified Organisms', 18 September, pp. 805–6.

—— (1996j) 'Boycott of Modified Soybean Escalates as Greenpeace Blocks Shipment', 13 November, p. 1005.

—— (1997a) 'Environment Council Reaches Mandate for Negotiations at Climate Change Meeting', 5 March, pp. 187–8.

—— (1997b) 'Commission Hopeful Climate Change Accord Will Help Boost Effort for CO_2/Energy Taxes', 5 March, pp. 191–2.

—— (1997c) 'Ministers Agree to 7.5 Percent Cut in Greenhouse Gas Emissions by 2005', 25 June, pp. 607–8.

—— (1997d) 'Senate Approves Resolution 95–0 Calling for Binding Controls on Developing Nations', 6 August, pp. 752–3.

—— (1997e) 'U.S., Japan, Other Nations Agree to Urge EU to Modify Proposal on Greenhouse Gas Cuts', 15 October, p. 951.

—— (1998a) 'Differences Remain Between U.S., EU on Defining Scope of Biosafety Protocol', 4 March, p. 188.

—— (1998b) 'Commission Fails to Get Votes to Start Legal Action Over Bt Maize Ban', 29 April, pp. 408–9.

—— (1998c) 'Austria Approves One of Toughest Laws in Europe on Genetically Modified Organisms', 29 April, pp. 409–10.

—— (1998d) 'Key Provisions of Biosafety Protocol Left for Final Negotiations in February 1999', 2 September, p. 847.

—— (1998e) 'Commission Takes First Step in Proceedings Against France Over GMO Maize Seed Ban', 14 October, pp. 1000–1.

—— (1998f) 'Coalition Government to Take More Cautious Approach to Genetically Modified Foodstuffs', 25 November, pp. 1176–7.

—— (1999a) 'Key Issue in Biosafety Protocol Talks is Coverage of Products or Just Organisms', 20 January, pp. 61–2.

—— (1999b) 'Talks on Biosafety Protocol Suspended; "Miami Group" Thwarts Compromise Accord', 3 March, pp. 177–9.

—— (1999c) 'U.S. State Department Nominee Cites Priority of Biosafety Protocol That Protects Trade', 9 June, p. 492.

—— (1999d) 'Informal Talks Seen to Reaffirm Commitment of All Parties to Agree on Biosafety Protocol', 29 September, pp. 785–6.

—— (1999e) 'State Department Environment Official Says Climate Change, Biodiversity Priorities', 8 December, pp. 996–7.

—— (2000a) 'DaimlerChrysler Leaves Industry Coalition Opposed to Legal Requirements to Cut GHGs', 19 January, p. 59.

—— (2000b) 'Texaco Leaves Industry Coalition, Maintains Opposition to Kyoto Pact', 15 March, pp. 244–5.

Inter Press Service (1997a) 'European Oil Companies React to Kyoto Deal', 15 December.

——— (1997b) 'American Petroleum Institute (API) Opposes Kyoto Agreement', 16 December.

IPCC (1995) 'IPCC Second Assessment: Climate Change 1995. A Report of the Intergovernmental Panel on Climate Change'. WMO/UNEP.

——— (2007) 'Climate Change 2007: The Physical Science Basis. Summary for Policymakers', Intergovernmental Panel on Climate Change, February.

Isaac, G.E., and W.A. Kerr (2007) 'The Biosafety Protocol and the WTO: Concert or Conflict?', in R. Falkner (ed.) *The International Politics of Genetically Modified Food: Diplomacy, Trade and Law* (Basingstoke: Palgrave Macmillan), pp. 195–212.

Jachtenfuchs, M. (1990) 'The European Community and the Protection of the Ozone Layer', *Journal of Common Market Studies* 28(3): 261–77.

Jaffe, A.B., S.R. Peterson, P.R. Portney and R.N. Stavins (1995) 'Environmental Regulation and the Competitiveness of U.S. Manufacturing: What Does the Evidence Tell Us?' *Journal of Economic Literature* 33, March: 132–63.

Jagers, S.C., and J. Stripple (2003) 'Climate Governance Beyond the State', *Global Governance* 9(3): 385–99.

James, C. (1999) 'Preview: Global Review of Commercialized Transgenic Crops: 1999', *ISAAA Briefs No. 12*, Ithaca, NY: ISAAA.

——— (2006) 'Global Status of Commercialized Biotech/GM Crops: 2006', *ISAAA Briefs No. 35*, Ithaca, NY: ISAAA.

Jessop, B. (1982) *The Capitalist State: Theory and Methods* (New York: New York University Press).

Johnston, J. (2003) 'CWB Asks Monsanto to Put the Brakes on Roundup Ready Wheat', AGWEB, 27 May.

Joly, P.-B., and S. Lemarié (1998) 'Industry Consolidation, Public Attitude, and the Future of Plant Biotechnology in Europe', *AgBioForum* 1(2): 85–90.

Jones, G. (2005) *Multinationals and Global Capitalism: From the Nineteenth to the Twenty-First Century* (Oxford: Oxford University Press).

Jordan, A. (1997) 'The Ozone Endgame: The Implementation of the Montreal Protocol in the UK', *CSERGE Working Paper*, GEC 97–16. Norwich.

Karas, J. (2004) 'Russia and the Kyoto Protocol: Political Challenges', Briefing Note, London: Chatham House.

Kathuri, C., E.T. Polastro and N. Mellor (1992) 'Biotechnology in an Uncommon Market', *Bio/Technology* 10(12): 1545–7.

Katzenstein, P.J. (1985) *Small States in World Markets: Industrial Policy in Europe* (Ithaca, NY: Cornell University Press).

Keck, M.E., and K. Sikkink (1998) *Activists Beyond Borders: Advocacy Networks in International Politics* (Ithaca, NY: Cornell University Press).

Keohane, R., and J.S. Nye, Jr. (eds) (1971) *Transnational Relations and World Politics* (Cambridge, MA: Harvard University Press).

Key, V.O. (1942) *Politics, Parties, and Pressure Groups* (New York: Thomas Y. Crowell).

King, J.L., N. Wilson and A. Naseem (2002) 'A Tale of Two Mergers: What We Can Learn from Agricultural Biotechnology Event Studies', *AgBioForum* 5(1): 14–19.

Koenig-Archibugi, M. (2004) 'Transnational Corporations and Public Accountability', *Government and Opposition* 39(2): 234–59.

Koester, V. (2002) 'The Biosafety Working Group (BSWG) Process: A Personal Account from the Chair', in C. Bail, R. Falkner and H. Marquard (eds) *The*

Cartagena Protocol on Biosafety: Reconciling Trade in Biotechnology with Environment and Development? (London: Earthscan), pp. 44–61.

Kolk, A. (2005) 'Corporate Social Responsibility in the Coffee Sector: The Dynamics of MNC Responses and Code Development', *European Management Journal* 23(2): 228–36.

Korten, D.C. (1995) *When Corporations Rule the World* (London: Earthscan).

Krimsky, S. (1982) *Genetic Alchemy: The Social History of the Recombinant DNA Controversy* (Cambridge, MA: MIT Press).

La Vina, A.G.M. (2002) 'A Mandate for a Biosafety Protocol', in C. Bail, R. Falkner and H. Marquard (eds) *The Cartagena Protocol on Biosafety: Reconciling Trade in Biotechnology with Environment and Development?* (London: Earthscan), pp. 34–43.

Large, A. (1976) 'The Spread of International Controls', *Wall Street Journal*, 22 November.

Leggett, J. (ed.) (1990) *Global Warming: The Greenpeace Report* (Oxford: Oxford University Press).

—— (ed.) (1996) *Climate Change and the Financial Sector: The Emerging Threat – The Solar Solution* (Munich: Gerling Akademie Verlag).

Levy, D.L. (1997) 'Business and International Environmental Treaties: Ozone Depletion and Climate Change', *California Management Review* 39(3): 54–71.

—— (2005) 'Business and the Evolution of the Climate Regime: The Dynamics of Corporate Strategies', in D.L. Levy and P.J. Newell (eds) *The Business of Global Environmental Governance* (Cambridge, MA: MIT Press), pp. 73–104.

—— (forthcoming) 'Political Contestation in Global Production Networks', *Academy of Management Review* 33(4).

Levy, D.L., and P.J. Newell (2000) 'Oceans Apart? Business Responses to the Environment in Europe and North America', *Environment* 42(9): 8–20.

—— (2005) 'A Neo-Gramscian Approach to Business in International Environmental Politics: An Interdisciplinary, Multilevel Framework', in D.L. Levy and P.J. Newell (eds) *The Business of Global Environmental Governance* (Cambridge, MA: MIT Press), pp. 47–69.

Levy, D.L., and A. Prakash (2003) 'Bargains Old and New: Multinational Corporations in Global Governance', *Business and Politics* 5(2): 131–50.

Lheureux, K., et al. (2003) 'Review of GMOs under Research and Development and in the Pipeline in Europe' (Brussels: European Commission Joint Research Centre).

Lindblom, C.E. (1977) *Politics and Markets: The World's Political-Economic Systems* (New York: Basic Books).

Litfin, K. (1994) *Ozone Discourses: Science and Politics in Global Environmental Cooperation* (New York: Columbia University Press).

Loader, R., and S. Henson (1998) 'A View of GMOs From the UK', *AgBioForum* 1(1): 31–4.

Lowi, T. (1964) 'American Business, Public Policy Case Studies and Political Theory', *World Politics* 16(4): 677–715.

Lukes, S. (1974) *Power: A Radical View* (London: Macmillan).

—— (2005) *Power: A Radical View*, 2nd edn (Basingstoke: Palgrave Macmillan).

Makhijani, A., and K.R. Gurney (1995) *Mending the Ozone Hole: Science, Technology, and Policy* (Cambridge, MA: MIT Press).

Mantegazzini, M.C. (1986) *The Environmental Risks from Biotechnology* (London: Frances Pinter).

Manufacturing Chemist (1988) 'Industry Gears up to Find CFC Alternatives', September.

Marquard, H. (2002) 'Scope', in C. Bail, R. Falkner and H. Marquard (eds) *The Cartagena Protocol on Biosafety: Reconciling Trade in Biotechnology with Environment and Development?* (London: Earthscan), pp. 289–98.

Martin, M. (2007) '5 Western States Announce Effort to Reduce Emissions', *San Francisco Chronicle*, 27 February.

Martinson, J. (1999) 'Monsanto Pays GM Price: Controversial Foods Division to Be Spun Off as Pharmaceuticals Groups Merge', *Guardian*, 21 December.

Mathews, J.T. (1997) 'Power Shift', *Foreign Affairs* 76(1): 50–66.

Max, A. (2000) 'Climate Talks Failure Leaves Businesses in the Dark and Disappointed', *Associate Press*, 2 December.

Maxwell, J.H., and S.L. Weiner (1993) 'Green Consciousness or Dollar Diplomacy? The British Response to the Threat of Ozone Depletion', *International Environmental Affairs* 5(1): 19–41.

May, C. (ed.) (2006) *Global Corporate Power* (Boulder, CO: Lynne Rienner).

McFarland, A.S. (2004) *Neopluralism: The Evolution of Political Process Theory* (Lawrence: University Press of Kansas).

Mesure, S. (2007) 'Tesco Follows M&S with Climate Change Move', *Independent*, 16 January.

Michaelowa, A. (1998) 'Impact of Interest Groups on EU Climate Policy', *European Environment* 8(5): 152–60.

Miliband, R. (1976) *The State in Capitalist Society* (London: Quartet Books).

Milmo, C. (1999) 'Sainsbury's Bans GM Food From Own Brand Range', *PA News*, 17 March.

Milner, H.V. (1988) *Resisting Protectionism: Global Industries and the Politics of International Trade* (Princeton, NJ: Princeton University Press).

Mizruchi, M.S. (1989) 'Similarity of Political Behavior Among Large American Corporations', *The American Journal of Sociology* 95(2): 401–24.

Molina, M., and F.S. Rowland (1974) 'Stratospheric Sink for Chlorofluoromethanes: Chlorine Atom-catalysed Destruction of Ozone', *Nature* (249): 810–12.

Monbiot, G. (2000) *Captive State: The Corporate Takeover of Britain* (Basingstoke: Palgrave Macmillan).

Moore, C., and A. Miller (1994) *Green Gold: Japan, Germany, the United States, and the Race for Environmental Technology* (Boston, MA: Beacon Press).

Mufson, S., and J. Eilperin (2006) 'Energy Firms Come to Terms With Climate Change', *Washington Post*, 25 November.

Munson, A. (1993) 'Genetically Manipulated Organisms: International Policy-Making and Implications', *International Affairs* 69(3): 497–517.

—— (1995) 'Should a Biosafety Protocol be Negotiated as Part of the Biodiversity Convention?' *Global Environmental Change* 5(1): 7–26.

Murphy, D.F., and J. Bendell (1997) *In the Company of Partners: Business, Environmental Groups and Sustainable Development Post-Rio* (Bristol: The Policy Press).

Nature Biotechnology (1999) 'GMO Panic Affects Drugs', 17(10): 939.

Nesmith, J. (2002) 'Industry-Linked Group Helped Push American Off Climate Panel', Cox News Service, 10 May.

Neumayer, E. (2001) *Greening Trade and Investment: Environmental Protection Without Protectionism* (London: Earthscan).

Newell, P. (2000a) *Climate for Change: Non-State Actors and the Global Politics of the Greenhouse* (Cambridge: Cambridge University Press).

—— (2000b) 'Environmental NGOs and Globalisation: The Governance of TNCs', in R. Cohen and S. Rai (eds) *Global Social Movements* (London: Athlone Press), pp. 117–34.

Newell, P., and M. Paterson (1998) 'A Climate for Business: Global Warming, the State and Capital', *Review of International Political Economy* 5(4): 679–703.

Niiler, E. (1999a) 'Monsanto Remains a Magnet for GM Opposition', *Nature Biotechnology* 17(9): 848.

—— (1999b) 'Terminator Technology Temporarily Terminated', *Nature Biotechnology* 17(11): 1054.

—— (2000a) 'Demise of the Life Science Company Begins', *Nature Biotechnology* 18(1): 14.

—— (2000b) 'Monsanto to Merge with P&U', *Nature Biotechnology* 18(2): 141.

Nogueira, A.H.V. (2002) 'Brazil', in C. Bail, R. Falkner and H. Marquard (eds) *The Cartagena Protocol on Biosafety: Reconciling Trade in Biotechnology with Environment and Development?* (London: Earthscan), pp. 129–37.

Nowell, G.P. (1996) 'International Relations Theories: Approaches to Business and the State', in R.W. Cox (ed.) *Business and the State in International Relations* (Boulder, CO: Westview Press), pp. 181–97.

Nunn, J. (2000) 'What Lies Behind the GM Label on UK Foods', *AgBioForum* 3(4): 250–4.

Oberthür, S. (1997) *Production and Consumption of Ozone-Depleting Substances, 1986–1995: The Data Reporting System under the Montreal Protocol* (Deutsche Gesellschaft für Technische Zusammenarbeit).

Oberthür, S., and T. Gehring (eds) (2006) *Institutional Interaction in Global Environmental Governance: Synergy and Conflict among International and EU Policies* (Cambridge, MA: MIT Press).

Oberthür, S., and H.E. Ott (1999) *The Kyoto Protocol: International Climate Policy for the 21st Century* (Berlin: Springer).

O'Brien, R., A.M. Goetz, J.A. Scholte and M. Williams (2000) *Contesting Global Governance: Multilateral Economic Institutions and Global Social Movements* (Cambridge: Cambridge University Press).

OECD (1976) 'Fluorocarbons: An Assessment of Worldwide Production, Use and Environmental Issues. First Interim Report', Environment Directorate, Paris: Organization for Economic Cooperation and Development.

—— (1981) 'Report on Chlorofluorocarbons', Environment Committee, Paris: Organization for Economic Cooperation and Development.

Ohmae, K. (1995) *The End of the Nation State: The Rise of Regional Economics* (London: HarperCollins).

Olson, R.D. (2005) 'Hard Red Spring Wheat at a Genetic Crossroad: Rural Prosperity or Corporate Hegemony?' in D.L. Kleinman, A.J. Kinchy and J. Handelsman (eds) *Controversies in Science and Technology. Vol. 1: From Maize to Menopause* (Madison: University of Wisconsin Press), pp. 150–68.

Oye, K.A., and J.H. Maxwell (1995) 'Self-Interest and Environmental Management', in R.O. Keohane and E. Ostrom (eds) *Local Commons and Global Interdependence: Heterogeneity and Cooperation in Two Domains* (Newbury Park: Sage).

Paarlberg, R. (1997) 'Earth in Abeyance: Explaining Weak Leadership in U.S. International Environmental Policy', in R.J. Lieber (ed.) *Eagle Adrift: American Foreign Policy at the End of the Century* (New York: Longman), pp 135–60.

Parsai, G. (2006) 'Exporters Seek Ban on Field Trials of GM Rice', *The Hindu*, 2 November.

Parson, E. (2003) *Protecting the Ozone Layer: Science and Strategy* (Oxford: Oxford University Press).

Paterson, M. (2001) 'Risky Business: Insurance Companies in Global Warming Politics', *Global Environmental Politics* 1(4): 18–42.

Paterson, W.E. (1991) 'Self-Regulation Under Pressure: Environmental Protection Policy in the Chemical Industry and the Response of the Sectoral Business Associations', in A. Martinelli (ed.) *International Markets and Global Firms: A Comparative Study of Organized Business in the Chemical Industry* (London: Sage), pp. 228–48.

Pattberg, P. (2007) *Private Institutions and Global Governance: The New Politics of Environmental Sustainability* (Cheltenham: Edward Elgar).

Patterson, L.A. (2000) 'Biotechnology Policy: Regulating Risks and Risking Regulation', in H. Wallace and W. Wallace (eds) *Policy-Making in the European Union* (Oxford: Oxford University Press), pp. 317–43.

Paugh, J., and J.C. Lafrance (1997) 'The U.S. Biotechnology Industry', U.S. Department of Commerce, Office of Technology Policy. Washington, DC, July.

Pollack, A. (1991) 'Moving Fast to Protect Ozone Layer', *New York Times*, 15 May.

Pollack, M.A., and G.C. Shaffer (2005) 'Biotechnology Policy', in H. Wallace, W. Wallace and M.A. Pollack (eds) *Policy-Making in the European Union*, 5th edn (Oxford: Oxford University Press), pp. 329–51.

Porter, G. (1992) 'The United States and the Biodiversity Convention', *EESI Papers on Environment and Development, No. 1*, Washington, DC: Environmental and Energy Study Institute.

Porter, M.E. (1985) *Competitive Advantage: Creating and Sustaining Superior Performance* (New York: The Free Press).

Porter, M., and C. van der Linde (1995) 'Green and Competitive: Ending the Stalemate', *Harvard Business Review* 73(5): 120–33.

PR Newswire (1992) 'Earth Summit: U.S. Business Endorses International Cooperation on Climate Change', 5 June.

—— (1997) 'Chrysler Statement in Response to Outcome of Global Climate Negotiations in Kyoto, Japan', 11 December.

—— (1999) 'Global Industry Coalition Reaffirms Its Support of a Successful Biosafety Protocol', 20 September.

Prakash, A. (2002) 'Beyond Seattle: Globalization, the Nonmarket Environment and Corporate Strategy', *Review of International Political Economy* 9(3): 513–37.

Prakash, A., and M. Potoski (2006) *The Voluntary Environmentalists: Green Clubs, ISO 14001 and Voluntary Environmental Regulations* (Cambridge: Cambridge University Press).

Pulver, S. (2002) 'Organizing Business: Industry NGOs in the Climate Debates', *Greener Management International* (39): 55–67.

—— (2007) 'Making Sense of Corporate Environmentalism: An Environmental Contestation Approach to Analyzing the Causes and Consequences of the

Climate Change Policy Split in the Oil Industry', *Organization and Environment* 20(1): 1–40.

Putnam, R.D. (1988) 'Diplomacy and Domestic Politics: The Logic of Two-Level Games', *International Organization* 42(3): 427–60.

Rabe, B.G. (2004) *Statehouse and Greenhouse: The Emerging Politics of American Climate Change Policy* (Washington, DC: Brookings Institution Press).

Raustiala, K. (1997) 'The Domestic Politics of Global Biodiversity Protection in the United Kingdom and the United States', in M.A. Schreurs and E. Economy (eds) *The Internationalization of Environmental Protection* (Cambridge: Cambridge University Press), pp. 42–73.

Reifschneider, L. (2002) 'Global Industry Coalition', in C. Bail, R. Falkner and H. Marquard (eds) *The Cartagena Protocol on Biosafety: Reconciling Trade in Biotechnology with Environment and Development?* (London: Earthscan), pp. 273–7.

Reiner, D.M. (2001) 'Climate Impasse: How The Hague Negotiations Failed', *Environment* 43(2): 36–43.

Reinhardt, F. (1989) 'Du Pont Freon® Products Division: Harvard Business School Case Study', Washington, DC: National Wildlife Federation.

Reuters (1999) 'Sainsbury Says Own-brand Ingredients GM-free', 19 July.

Rifkin, J. (1998) *The Biotech Century: Harnessing the Gene and Remaking the World* (New York: Jeremy P. Tarcher/Putnam).

Risse-Kappen, T. (ed.) (1995) *Bringing Transnational Relations Back In: Non-State Actors, Domestic Structures and International Institutions* (Cambridge: Cambridge University Press).

Roan, S. (1989) *Ozone Crisis: The 15-Year Evolution of a Sudden Global Emergency* (New York: John Wiley & Sons).

Rogowski, R. (1989) *Commerce and Coalitions: How Trade Affects Domestic Political Alignments* (Princeton, NJ: Princeton University Press).

Rosecrance, R.N. (1999) *The Rise of the Virtual State: Wealth and Power in the Coming Century* (New York: Basic Books).

Rowland, F.S., and M. Molina (1976) 'Stratospheric Formation and Photolysis of Chlorine Nitrate', *Journal of Physical Chemistry* 80(24): 2711–13.

Rowlands, I.H. (1995) *The Politics of Global Atmospheric Change* (Manchester: Manchester University Press).

—— (2000) 'Beauty and the Beast? BP's and Exxon's Positions on Global Climate Change', *Environmental and Planning* 18: 339–54.

Salt, J. (1998) 'Kyoto and the Insurance Industry: An Insider's Perspective', *Environmental Politics* 7(2): 161–5.

Samper, C. (2002) 'The Extraordinary Meeting of the Conference of the Parties (ExCOP)', in C. Bail, R. Falkner and H. Marquard (eds) *The Cartagena Protocol on Biosafety: Reconciling Trade in Biotechnology with Environment and Development?* (London: Earthscan), pp. 62–75.

Sawin, J.L. (2004) 'Meanstreaming Renewable Energy in the 21st Century', Worldwatch Paper #169, Worldwatch Institute, May.

Schattschneider, R.E. (1935) *Politics, Pressure and the Tariff* (New York: Prentice-Hall).

Schmidheiny, S., and BCSD (1992) *Changing Course: A Global Business Perspective on Development and the Environment* (Cambridge, MA: MIT Press).

Schmidt, V.A. (1995) 'The New World Order, Incorporated: The Rise of Business and the Decline of the Nation-State', *Dædalus* 124(2): 75–106.

Schurman, R. (2004) 'Fighting "Frankenfoods": Industry Opportunity Structures and the Efficacy of the Anti-Biotech Movement in Western Europe', *Social Problems* 51(2): 243–68.

Selin, H., and S.D. VanDeveer (2007) 'Political Science and Prediction: What's Next for U.S. Climate Change Policy?' *Review of Policy Research* 24(1): 1–27.

Sell, S.K. (2003) *Private Power, Public Law: The Globalization of Intellectual Property Rights* (Cambridge: Cambridge University Press).

Shackley, S., and J. Hodgson (1991) 'Biotechnology Regulation in Europe', *Bio/Technology* 9(11): 1056–61.

Shanley, P., A.R. Pierce, S.A. Laird and A. Guillén (eds) (2002) *Tapping the Green Market: Certification and Management of Non-Timber Forest Products* (London: Earthscan).

Shapiro, I. (2006) 'On the Second Edition of Lukes' Third Face', *Political Studies Review* 4(2): 146–55.

Sheingate, A.D. (2006) 'Promotion Versus Precaution: The Evolution of Biotechnology Policy in the United States', *British Journal of Political Science* 36(2): 243–68.

Shimoda, S. (1994) 'Agbiotech Will Vertically Integrate Agribusiness', *Bio/Technology* 12(11): 1062–4.

Skidmore, D. (1995) 'The Business of International Politics', *Mershon International Studies Review* 39(2): 246–54.

Skidmore-Hess, D. (1996) 'Business Conflict and Theories of the State', in R.W. Cox (ed.) *Business and the State in International Relations* (Boulder, CO: Westview Press), pp. 199–216.

Skjærseth, J.B. (1994) 'The Climate Policy of the EC: Too Hot to Handle?' *Journal of Common Market Studies* 32(1): 25–45.

Skjærseth, J.B., and T. Skodvin (2003) *Climate Change and the Oil Industry: Common Problem, Varying Strategies* (Manchester: Manchester University Press).

Skogstad, G. (2003) 'Legitimacy and/or Policy Effectiveness? Network Governance and GMO Regulation in the European Union', *Journal of European Public Policy* 10(3): 321–38.

Smelser, N.J., and R. Swedberg (2005) 'The Handbook of Economic Sociology' (Princeton, NJ: Princeton University Press).

Smith, M.J. (1993) *Pressure, Power and Policy: State Autonomy and Policy Networks in Britain and the United States* (New York: Harvester Wheatsheaf).

Somheil, T. (1996) 'Time to Optimize', *Appliance*, October.

Spar, D.L. (2001) *Pirates, Prophets and Pioneers: Business and Politics Along the Technological Frontier* (New York: Random House).

Steger, M.B. (2003) *Globalization: A Very Short Introduction* (Oxford: Oxford University Press).

Stern, N. (2007) *The Economics of Climate Change: The Stern Review* (Cambridge: Cambridge University Press).

Stopford, J., and S. Strange (1991) *Rival States, Rival Firms: Competition for World Market Shares* (Cambridge: Cambridge University Press).

Strange, S. (1988) *States and Markets: An Introduction to International Political Economy* (London: Pinter).

—— (1996) *The Retreat of the State: The Diffusion of Power in the World Economy* (Cambridge: Cambridge University Press).

Streeck, W. (1992) *Social Institutions and Economic Performance: Studies of Industrial Relations in Advanced Capitalist Economies* (London: Sage).

Süddeutsche Zeitung (1988) 'Hoechst will Ozonschicht schützen', 7 December.

Tapper, R. (2002) 'Environment Business and Development', in C. Bail, R. Falkner and H. Marquard (eds) *The Cartagena Protocol on Biosafety: Reconciling Trade in Biotechnology with Environment and Development?* (London: RIIA/Earthscan), pp. 268–72.

Tarrow, S. (2005) *The New Transnational Activism* (Cambridge: Cambridge University Press).

Torgersen, H., et al. (2002) 'Promise, Problems and Proxies: Twenty-Five Years of Debate and Regulation in Europe', in M.W. Bauer and G. Gaskell (eds) *Biotechnology – The Making of a Global Controversy* (Cambridge: Cambridge University Press), pp. 21–94.

Truman, D.B. (1951) *The Governmental Process* (New York: Knopf).

Tulder, R.V., and A. Kolk (2001) 'Multinationality and Corporate Ethics: Codes of Conduct in the Sporting Goods Industry', *Journal of International Business Studies* 32(2): 267–83.

UNEP (1989a) 'Aerosols, Sterilants and Miscellaneous Uses of CFCs', Report, Technical Options Committee on Aerosols, Sterilants and Miscellaneous Uses, United Nations Environment Programme, 30 June.

—— (1989b) 'Synthesis Report', UNEP/OzL.Pro.WG.II(1)/4. United Nations Environment Programme, 13 November.

—— (1989c) 'Technical Progress on Protecting the Ozone Layer', Report of the Technology Review Panel, United Nations Environment Programme, 30 June.

—— (1995) 'Statement of Environmental Commitment by the Insurance Industry', Geneva.

US House of Representatives (1975) 'Fluorocarbons – Impact on Health and Environment', Hearing, Committee on Interstate and Foreign Commerce, Subcommittee on Public Health and Environment, 11–12 December 1974. Washington, DC: Governmental Printing Office.

—— (1981) 'EPA Rulemaking on Chlorofluorocarbons (CFCs) and Its Impact on Small Business', Hearing, Committee on Small Business, Subcommittee on Antitrust and Restraint of Trade Activities Affecting Small Business, 15 July 1981. Washington, DC: Government Printing Office.

—— (1987) 'Ozone Layer Depletion', Hearing, Committee on Energy and Commerce, Subcommittee on Health and Environment, 9 March 1987. Washington, DC: Government Printing Office.

US Senate (1982) 'Nominations of Anne M. Gorsuch and John W. Hernandez, Jr.', Hearings, Committee on Environment and Public Works, 1 and 4 May 1982. Washington, DC: Government Printing Office.

—— (1986) 'Ozone Depletion, the Greenhouse Effect, and Climate Change', Hearings, Committee on Environment and Public Works, Subcommittee on Environmental Pollution, 10 and 11 June 1986. Washington, DC: Government Printing Office.

—— (1994) 'Stratospheric Ozone Depletion', Hearing, Committee on Governmental Affairs, Ad Hoc Subcommittee on Consumer and Environmental Issues, 17 December 1991. Washington, DC: Government Printing Office.

Van der Woerd, F. (2005) 'The Chemical Industry's Response to Climate Change', in K. Begg, F. Van der Woerd and D. Levy (eds) *The Business of Climate Change: Corporate Responses to Kyoto* (Sheffield: Greenleaf Publishing), pp. 189–95.

Vernon, R. (1998) *In the Hurricane's Eye: The Troubled Prospects of Multinational Enterprises* (Cambridge, MA: Harvard University Press).

Victor, D.G., and J.C. House (2006) 'BP's Emissions Trading System', *Energy Policy* 34(15): 2100–12.

Vig, N.J. (1990) 'Presidential Leadership: From the Reagan to the Bush Administration', in N.J. Vig and M.E. Kraft (eds) *Environmental Policy in the 1990s: Toward a New Agenda* (Washington, DC: Congressional Quarterly Press), pp. 33–58.

Vogel, D. (1988) *Fluctuating Fortunes: The Political Power of Business in America* (New York: Basic Books).

—— (1995) *Trading Up: Consumer and Environmental Regulation in a Global Economy* (Cambridge, MA: Harvard University Press).

—— (2003) 'The Hare and the Tortoise Revisited: The New Politics of Consumer and Environmental Regulation in Europe', *British Journal of Political Science* 33(4): 557–80.

—— (2005) *The Market for Virtue: The Potential and Limits of Corporate Social Responsibility* (Washington, DC: Brookings Institution Press).

Vogler, J., and C. Bretherton (2006) 'The European Union as a Protagonist to the United States on Climate Change', *International Studies Perspectives* 7(1): 1–22.

Vrolijk, C. (2001) 'COP-6 Collapse or "to be continued...?"', *International Affairs* 77(1): 163–9.

Walgate, R. (1990) *Miracle or Menace? Biotechnology and the Third World* (London: The Panos Institute).

Wall Street Journal (1989) 'Hoechst AG Wants to Have CFC Replacement by 1995', 26 October.

Waltz, K.N. (2000) 'Globalization and American Power', *The National Interest* (59): 46–56.

Wapner, P. (1996) *Environmental Activism and World Civic Politics* (Albany: State University of New York Press).

Weart, S.R. (2003) *The Discovery of Global Warming* (Cambridge, MA: Harvard University Press).

Weber, M. (1964) *The Theory of Social and Economic Organization*, translated by A.M. Henderson and Talcott Parsons (New York: Free Press).

Weiss, L. (1999) 'Globalisation and National Governance: Antinomy or Interdependence?' *Review of International Studies* 25(5): 59–88.

Weiss, R. (1999) 'British Revolt Grows Over "Genetic" Foods', *Washington Post*, 29 April.

Wendt, A. (1992) 'Anarchy is What States Make of it: The Social Construction of Power Politics', *International Organization* 46(2): 391–425.

Wheeler, D., H. Fabig and R. Boele (2002) 'Paradoxes and Dilemmas for Stakeholder Responsive Firms in Extractive Sectors: Lessons from the Case of Shell and the Ogoni', *Journal of Business Ethics* 39(3): 297–318.

Williams, D. (1997) 'Green Groups Say Kyoto Protocol Endangers Climate', *Agence France Presse*, 11 December.

Williams, M. (2001) 'In Search of Global Standards: The Political Economy of Trade and the Environment', in D. Stevis and V.J. Assetto (eds) *The International Political Economy of the Environment: Critical Perspectives* (Boulder, CO: Lynne Rienner), pp. 39–61.

Wilson, G.K. (2006) 'Thirty Years of Business and Politics', in D. Coen and W. Grant (eds) *Business and Government: Methods and Practice* (Opladen: Barbara Budrich), pp. 33–50.

Wisner, R.N. (2003) 'Market Risks of Genetically Modified Wheat: The potential short-term impacts of GMO Spring Wheat introduction on U.S. wheat export markets and prices', Iowa State University, 30 October.

—— (2004) 'Round-Up Ready® Spring Wheat: Its potential short-term impacts on U.S. wheat export markets and prices' *Economic Staff Report*, Iowa State University Department of Economics, Ames, Iowa, 1 July.

Woll, C. (2007) 'Leading the Dance? Power and Political Resources of Business Lobbyists', *Journal of Public Policy* 27(1): 57–78.

World Energy Council (2007) 'The Energy Industry Unveils Its Blueprint for Tackling Climate Change', WEC Statement, London, March.

Wright, C., and A. Rwabizambuga (2006) 'Institutional Pressures, Corporate Reputation, and Voluntary Codes of Conduct: An Examination of the Equator Principles', *Business and Society Review* 111(1): 89–117.

Wright, S. (1994) *Molecular Politics: Developing American and British Regulatory Policy for Genetic Engineering, 1972–1982* (Chicago: University of Chicago Press).

Zapfel, P., and M. Vainio (2002) 'Pathways to European Greenhouse Gas Emissions Trading History and Misconceptions', Nota di Lavoro 85.2002, October, Fondazione Eni Enrico Mattei.

Zedan, H. (2002) 'The Road to the Biosafety Protocol', in C. Bail, R. Falkner and H. Marquard (eds) *The Cartagena Protocol on Biosafety: Reconciling Trade in Biotechnology with Environment and Development?* (London: Earthscan), pp. 23–33.

Index